FOR THE MONTH OF

	Ledger Folio.	General Charges.	Native Labour Commission.	Native Wages.	Wages.	Persons.	General.
Forward							

CENTRAL ADMINISTRATION MINE CASH

Date.		Ledger Folio.	Sundry Persons.	East Rand Proprietary Mines.	Native Advances.	Chinese Advances.	Employees' Accident Assurance Claims Account.	General
	Forward							

ACCOUNTS AND DRAWINGS

THE AFRICA LIST

ACCOUNTS AND DRAWINGS FROM UNDERGROUND

EAST RAND PROPRIETARY MINES CASH BOOK, 1906

William Kentridge

Rosalind C. Morris

Seagull Books

CALCUTTA LONDON NEW YORK

Series Editor: Rosalind C. Morris

Seagull Books, 2015

ISBN 978 0 8574 2 205 7

British Library Cataloguing-in-Publication Data
A catalogue record for this book is available from the British Library

Designed by Sunandini Banerjee, Seagull Books

Printed and bound by
Hyam Enterprises, Calcutta, India

DRAWINGS

WILLIAM KENTRIDGE

CENTRAL ADMINISTRATION MINE CASH

Date.		Ledger Folio.	Sundry Persons.	East Rand Proprietary Mines.	Native Advances.	Chinese Advances.	Employees' Accident Assurance Claims Account.	General
	Forward							

FOR THE MONTH OF ______________ 190

Date.		Ledger Folio.	General Charges.	Native Labour Commission.	Native Wages.	Chinese Wages.	Sundry Persons.	General.
	Forward							

CENTRAL ADMINISTRATION MINE CASH

Date		Ledger Folio.	Sundry Persons.	East Rand Proprietary Mines.	Native Advances.	Chinese Advances.	Employees' Accident Assurance Claims Account.	General.
1906	Forward		488 7 1	805 12 10	234 13 4	1047 13 4	112 2 5	21890 - 1
Feb 24	Native Return Home Expenses	1537						27 2 6
	natives at Driefontein							
	natives at Driefontein				29 10 -			
	" " New Comet				26 12 -			
	" " Cason (transferred to Comet)				1 5 6			
	" " "				13 1 -			
	B.C.A Remittance a/c	739						21 - -
	natives at Cason							
	natives at Cason (transferred to Comet)				3 11 6			
	Chinese at New Comet					2 10 -		
	A Roodt a/c Stand Rent			2 - -				
	Lester A.L.	555	5 -					
	Waggon & trolley a/c							
	Acutt R.C.	20	8 - 6					
	Coloured Hospl Charges Dec. 10 6							
	do. do Jany. 4 15 -							
	European Labour " 10 -							
	Coal a/c " 2 5 -							
	Deceased Chinese Wages Suspense a/c	683						1 16 8
	Coolie at Angelo 19 9							
	" " " 16 11							
	Smart T.	897	10 -					
	Waggon & trolley a/c							
	Ferguson F.L.	327	2 6					
	Waggon & trolley a/c							
	Cain E. Angelo Compound	175	37 4 1					
	Amount received from Chamber of Mines Labour Importation Agency as per their Statement dated 19th Feby 1906							
	Coal Cash Sales a/c	138	42 4 -					
	Recd from Sundry Persons							
	Recd from R. Legg for key of room not returned			10 -				
	Turpin R.G.	870	2 9 -					
	Coal a/c 7/6 Store No 4 £2:1:6 Dec.							
	Chamber of Mines L.I. Agency Ltd	154	18 -					
	Amount of wages due to Coolie No 1052 New Comet							
	Koppel Ltd Arthur	504	8 17 7					
	Native Labour supplied Jany							
	Forward		588 17 9	808 2 10	308 13 4	1050 3 4	112 2 5	21939 19 3

FOR THE MONTH OF February 1906

Date.		Ledger Folio.	General Charges.	Native Labour Commission.	Native Wages.	Chinese Wages.	Sundry Persons.	General.
1	Sowter & Davidson's a/c for duplicate keys for safes &c		2 6 .					
	Passes for 5,725 Natives			572 10 .				
	Registration 60 Natives (12 mos)			3				
	Passes 60 " "			6				
	Native Compound Chgs Driefontein	1478						3
	Wages to Kaffir woman for brewing beer							
	Chinese Compound Chgs Angelo	33[illegible]						5 5
	Expenses in connection with case Rex v Ashurst.							
	Messenger's expenses to town with Reports (Prop of 6/6 2/2)		. 2 2					
	Hercules Co. Ltd	495					2 2	
	Boksburg Gold Mines Ltd.	119					2 2	
	Prop of Messengers expenses to town with Reports 6/6							
	Employees' Accident Assn Claims A/c	761.						
	Paid the following in full settlement of their claims	do						242 18 5
	Stone F.A. £3 : 15 . –							
	Wickham J. 156 : 10 : –							
	Heltne L.S. 5 : – : –							
	Hocking W.H. 77 : 13 : 5							
	Main Store A/c.	1434						100
	Petty Cash							
	Foster W	323					39 2 2	
	As per Journal Voucher R 597/48							
	Chinese Repatriation A/c	561.						1 5
	Train fares for Coolies							
	Chinese Defaulters A/c	573						1 1
	Expenses of Police conducting Coolies back to Mine							
	Passes for 17 Natives			1 14				
	Regn. 13 "			13				
	Police boys expenses to W.N.L. Assocn & Portuguese Curator			10				
	Messenger's expenses to Johannesburg with Gold Output for Chamber of Mines		5					
	Job 351/60 /60	1239						5
	One Revenue Stamp for application to Registrar of Mining Rights for Surface Rights for Railway Extension							
	Messengers expenses to Town with Surface Plan of Property		6 6					
	Forward		2 19 8	584 7			39 6 6	353 14 5

CENTRAL ADMINISTRATION MINE CASH

Date.	Particulars	Ledger Folio.	Sundry Persons. £	s	d	East Rand Proprietary Mines. £	s	d	Native Advances. £	s	d	Chinese Advances. £	s	d	Employees' Accident Assurance Claims Account. £	s	d	General. £	s	d
	Forward		996	13	6	1724	17	1	430	7	9	1315	19	4	763	2	10	97548	16	
30	Bates E.	103	1	15																
	Berry T	67	1	15																
	Chubb J. S.	176	3	8																
	Chadwick J	146	1	1																
	Edwards F. W.	246	3	19	6															
	Ferguson F. C.	316	1	10																
	Galway R.	356	2	15																
	Geddes W. L.	355	1	15																
	Giddey C.	377	2	13	6															
	Heath F. A.	464	5	16																
	Hoskyn C. H.	465	2	2	6															
	Kirk J	511		17	6															
	Liddelow C. C. W.	542		18	6															
	Lindhardt J. B.	534		18	6															
	Moore J. H.	581	2	3																
	McColl D. A.	577	1	15																
	Pagan D	724	1	9																
	Pearce A. J.	725	1	15																
	Robson J. N.	772	2	13	6															
	Sawyer R. E	812	1	10																
	Strang J	840	1	14																
	Strang J.	840	1	14																
	Taylor H. J	872	12																	
	Wademan A	980	1	19																
	J Peters on a/c of Stand Rent					4														
	Nicholls J. G.	623	1	16																
	Amounts collected as per April Pay Sheets.																			
	Driefontein Consolidated Mines Ltd Stand Rent due by M. Thompson for April.	588																a/c	7	6
	New Comet G. M. Co Ltd. Stand Rent due by T. Hill for April	561																a/c	7	6
	Angelo Gold Mines Ltd Stand Rent due by J. T. Davies for April	17																a/c	7	6
	J. T. Davies £56-16- M. Thompson 40- T. J. Hill 31-14- M. Faull 5- M. Faull 3-10- being rent of Angelo, D'fontein, New Comet and D'fontein Hill Boarding Houses and cottage D'fontein Hill for April.					137														
	Forward		1058	7	"	1865	17	1	430	7	9	1315	19	4	763	2	10	97549	18	6

FOR THE MONTH OF *April* 1906

Date.		Ledger Folio.	General Charges.	Native Labour Commission.	Native Wages.	Chinese Wages.	Sundry Persons.	General.
	Forward		1 · 9	571 4 6	5370 4 8	5914 - 10	2105 4 3	401 6 3
16	Exchange on £8000.		5 . .					
	Stamps		1					
	Angelo Gold Mines Ltd.	16						7 9
	G. Cockburn's expenses taking gold to town.							
	New Comet G. M. Co. Ltd.	1560						1 2 9
	G. Cockburn's expenses taking gold to town.							
	Messengers expenses to town with Weekly Reports		6 9					
	Chinese Compound Chgs. Angelo	385						2 3
	Telegram to Mansel & Davies re. unsatisfactory state of mealies							
	Regn 24 Natives (3 mos)			1 4 .				
	Passes 30 "			3				
	Transfers 3 "			3				
	Passes 10 "			1				
	Regn 7 " (3 mos)			7				
	Passes 12 "			1 4				
	Regn 11 " (6 mos)			11				
	Passes 30 "			3				
	Transfers 2 "			2				
	Regn 21 " (3 mos)			1 1				
	Stamps		1					
	Employees Accident Assn. Claims A/c	761						49 7
	Paid the following in full settlement of their claims.							
	Turnbull A. £5 : 15 : -							
	Hanson A. 7 : 5 : -							
	Bell H. D. 12 : - : -							
	Littlejohn W. 10 : - : -							
	Shave W. 5 : 10 : -							
	Meldrum R. 2 : - : -							
	Hooper J. 6 : 17 : 6							
	Tyser A.	898					18 17 7	
	As per J. V. 34/61.							
	Main Store A/c.	1435						175
	Further on a/c of Petty Cash.							
	Chinese Defaulters A/c.	573						1 10
	Expenses incurred in recovering Coolies from Gaol.							
17	Coolie at Cason deceased.					13 5		
	(Blair W.) Cason Gold Mines Ltd.	388					~~237 5~~	237 5
	As per Journal Voucher paid W. Blair							
	A. J. Fowles expenses to town with Reports		6					
	Forward		8 13 6	582 16 6	5370 4 8	5914 14 3	2124 1 10	866

CENTRAL ADMINISTRATION MINE CASH

Date.		Ledger Folio.	Sundry Persons.	East Rand Proprietary Mines.	Native Advances.	Chinese Advances.	Employees' Accident Assurance Claims Account.	General
	Forward		874 1 6	86 12 1	376 5 3	992 19 4	560 7 10	97383 9 7
25	Wagon & Trolley Cash Sales A/c	996	3 2 6					
	Recd from Sundry Persons							
	Coolies at Angelo 8th arrivals					166 10		
	Chinese Defaulters A/c	573						1 15 5
	Coolies at Angelo.							
	Chinese Remittance A/c	533						63 15
	Coolies at Angelo							
	Natives Return Home A/c	1527						14 17 6
	Natives at Driefontein							
	Natives at Driefontein				54 2 6			
	B.C.A. Remittance A/c	709						6
	Natives at Comet							
	Natives Return Home A/c	1527						17 6
	Natives at Comet							
	B.C.A. Remittance A/c							~~8~~
	Deceased Chinese Suspense A/c	683						1 1 2
	Coolie at Cason							
27	Chinese Defaulters A/c	573						1 10
	Coolies at New Comet							
	Coolies at Angelo					156 10		
	Chinese Defaulters A/c	573						14 10
	Coolies at Angelo							
	Chinese Remittance A/c	533						74 15
	Coolies at Angelo							
	Coal Cash Sales A/c	438	43 1 6					
	Recd from Sundry Persons							
	Jansen A.D.	471	7 6					
	Coal A/c March							
	Dickinson H.E.	210	12 6					
	Workshops Store No. 4 April							
	Forder L.E.	311	9 3					
	Main Store A/c April							
	A. Roodt Stand Rent Mar & April			2				
	W.J. Stanford rent for April			3 10				
	T.W. Stocker rent for April			6 14				
	Forward		921 14 9	98 16 1	430 7 9	1315 19 4	560 7 10	97548 16 –

FOR THE MONTH OF April 1906

Date		Ledger Folio.	General Charges.	Native Labour Commission.	Native Wages.	Chinese Wages.	Sundry Persons.	General.
	Forward		24 7 3	587 15 "	7632 4 8	9596 8 11	2130 16 10	902 2 "
20	Natives at New Comet				1744 12 8			
	" " Driefontein Hill				1 3			
	" " Driefontein				2 2			
	" " "				1 15			
	" " "				3			
	" " Driefontein Hill				11 8 4			
	" " " "				20 15			
	New Comet G. M. Co. Ltd G.W. Solomon's expenses to town with gold	1561						18
	Driefontein Consolidated Mines Ltd G.W. Solomon's expenses to town with gold	588						6 3
	Miller & Baptie As per Journal Voucher	574.					250	
	Convict Labour A/c Bonus to Warders Mess	479						20
	Chamber of Mines L. I. Agency Ltd As per Journal Voucher R 31/91	477.					8 7 6	
	Ford Sergt. W. J. As per Journal Voucher R 27/90	301.					10	
	Kleinfontein Central Administration As per Journal Voucher R 28/90	508.					1 10	
	Taylor P. E. As per Journal Voucher	879.					71 2 6	
	Chamber of Mines L. I. Agency As per Journal Voucher R 33/92	477.					8 9 10	
	Native Commissioner As per Journal Voucher R 36/92	637.					20	
	Welch A. A As per Journal Voucher	964.					8	
	Godwin & Anderson (As per Journal Voucher) On a/c of Contract	371.					150	
	Stamps		1					
	Boarding House Licences A/c Licence for D & A. Boarding House for three months ending 31st June 1906	1334						2 1 3
	Olders expenses to town with Weekly Reports		5					
	Older's expenses to town with Annual Reports		7					
	Employees Accident Assn Claims A/c Paid John Hunter in full settlement of his claim	761						5 15
	Forward		25 19 3	587 15 "	9417 " 8	9596 8 11	2648 16 8	931 2 6

CENTRAL ADMINISTRATION MINE CASH

Date		Ledger Folio.	Sundry Persons.	East Rand Proprietary Mines.	Native Advances.	Chinese Advances.	Employees' Accident Assurance Claims Account.	General
	Forward		✓834 19 4	2331 19 4	390 12 10	1050 3 4	142 17 5	✓92954 5
28	Lamb W.L.	535	1 6 6					
	Macdonald J 97 M. Qrs	588	1 14 6					
	Mackay A.N.	595	2 6					
	Opie D.	665	10					
	Powell H.V.	719	17 3					
	Smith C.A.	808	3 6					
	Sweet J.W.	811	21 11 2					
	Sansom F.L.	817	7 6					
	Solomon G.W.	826	7 6					
	Sawers A.A.	839	9					
	Smith W.R.	846	12 6					
	Smith R.B.	859	1 2					
	Taylor H.J.	872	15					
	Upton G.	919	2 6					
	Wyse T.A.	972	3					
	Watermeyer O.	984	7 6					
	A/cs due 31st Dec 1905 collected as per Pay Sheets							
	Boarding House Com A/c	737 ✓						35 15
	M. Morrison 1% on £999.10.6 (15) £10:–:–							
	M. Thompson ½% 1106:11:10 5:10:–							
	J.T. Davies ½% 2023:9:11 10:2:6							
	M. Faull ½% 309:5:6 1:10:–							
	T.J. Hill ½% 1920:5:8							
	less overcharge £1 last month 8:12:6							
	Davies J.T. £40:–:–							
	Thompson M. 56:16:–							
	Hill T.J. 31:14:–							
	Faull M 5:–:–							
	Faull M 3:10:–			137 ✓				
	Rent for Feby 1905.							
	Boarding House Licences A/c	1334 ✓						1 17 6
	Recd from the following props of Licences for Feby.							
	M. Thompson £:12:6							
	J.T. Davies 12:6							
	T.J. Hill 12:6							
	New Comet G.M. Co. Ltd.	1560 ✓						a/c 7 6
	Recd from T.J. Hill stand rent for month of February							
	Angelo Gold Mines Ltd	15 ✓						a/c 7 6
	Recd from J.T. Davies stand rent for month of February							
	Driefontein Consolidated Mines Ltd	587 ✓						a/c 7 6
	Recd from M. Thompson stand rent for month of February							
	Forward		✓867 12 11	2468 19 4	390 12 10	1050 3 4	142 17 5	✓92992 15 5

FOR THE MONTH OF *April* 1906

Date		Ledger Folio.	General Charges.	Native Labour Commission.	Native Wages.	Chinese Wages.	Sundry Persons.	General.
	Forward		- - -	4 17 6	247 2 3	5 15 4	1670 2 5	28 18 10
7	Natives at Driefontein repatriated				31 12 4			
	Olders expenses to town with Gold Output Returns.		7 6					
	Main Store A/c. On a/c of Petty Cash.	1433						100
	Employees Accident Assⁿ Claims A/c. Paid the following in full settlement: Thomas W. £39-11-8; Sutton J. W. H. 2:5:10; Olsen C. 4:15:-; Wilkinson A. 5:-:-	761						51 12 6
	Driefontein Consolidated Mines Ltd. G.W. Solomon's expenses to town with gold.	588						1 5 6
	New Comet G. M. Co. Ltd. G.W. Solomon's expenses to town with gold.	1560						12 6
	Angelo Gold Mines Ltd. G.W. Solomon's expenses to town with gold.	16						1 14
	Unclaimed European Wages. Paid Robert Tait wages due to him C. B. Nº 1 fol. 9.	1987						13 6
	East Rand Prop. Mines Ltd. Paid F. J. Aston, Salary from 1st to 18th April at £40 per mo. (debit Secretarial Fees)	737						23 4 6
	Farrar Bros. Messengers expenses to Johannesburg with Special Assays in connection with Rhodesian Samples	~~325~~ 33					7	
	Messengers expenses to town with letters for Mr Hellmann's signature		6 9					
	Messengers expenses to town with specimen of blood of Coolie Nº 5119 for analysis		6 6					
	Messengers expenses to Johannesburg with letter for Board Meeting re small footage developed each week.							
	Natives Return Home A/c. 55 Railway fares of boys returning home.	1537						1 12 3
	Forward		1 . 9	4 17 6	278 14 7	5 15 4	1670 9 5	222 6 1

CENTRAL ADMINISTRATION MINE CASH

DATE.		Ledger Folio.	Sundry Persons.	East Rand Proprietary Mines.	Native Advances.	Chinese Advances.	Employees' Accident Assurance Claims Account.	General
	Forward							

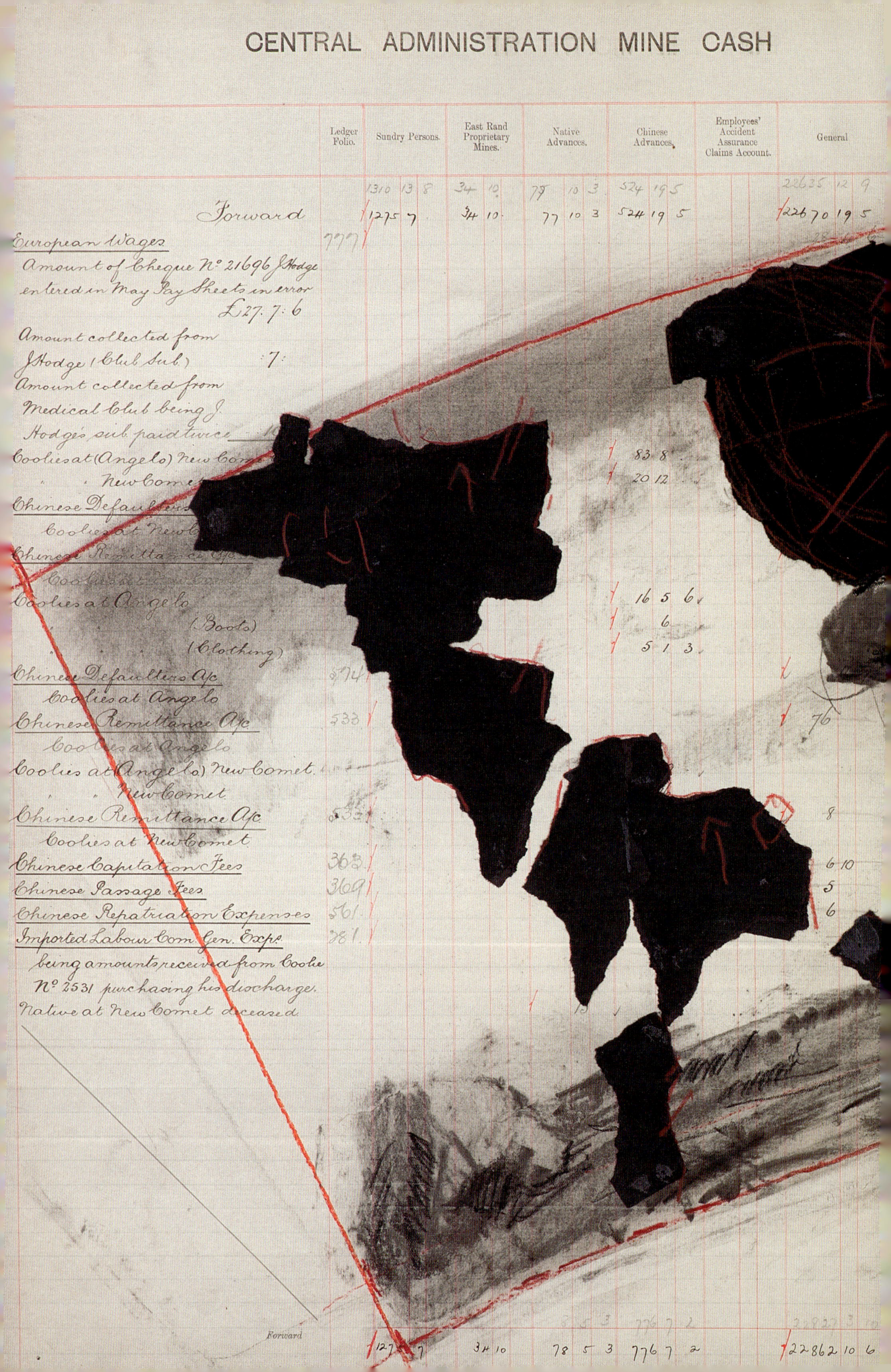
CENTRAL ADMINISTRATION MINE CASH
Ledger Folio.
Sundry Persons.
East Rand Proprietary Mines.
Native Advances.
Chinese Advances.
Employees' Accident Assurance Claims Account.
General
Forward
European Wages
Amount of Cheque No 21696 J Hodge entered in May Pay Sheets in error
£27. 7. 6
Amount collected from J Hodge (Club Sub)
Amount collected from Medical Club being J. Hodge's sub paid twice
Coolies at (Angelo) New Comet
New Comet
Chinese Defaulters A/c
Chinese Remittance A/c
Coolies at Angelo
(Boots)
(Clothing)
Chinese Defaulters A/c
Coolies at Angelo
Chinese Remittance A/c
Coolies at Angelo
Coolies at (Angelo) New Comet.
New Comet
Chinese Remittance A/c
Coolies at New Comet
Chinese Capitation Fees
Chinese Passage Fees
Chinese Repatriation Expenses
Imported Labour Com Gen. Exps
being amounts received from Coolie No 2531 purchasing his discharge.
Native at New Comet deceased
Forward

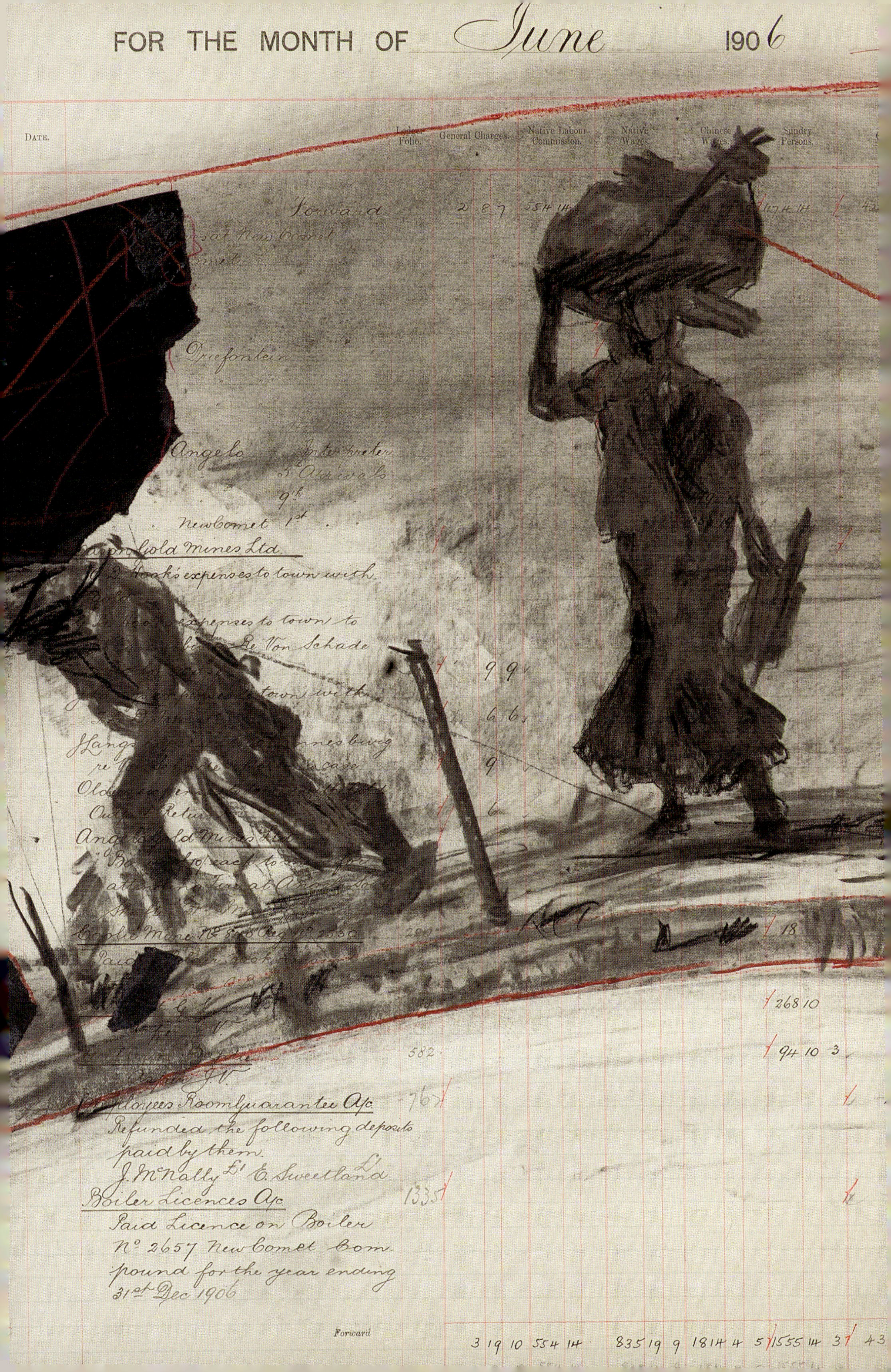
FOR THE MONTH OF June 1906
Date.
Ledger Folio.
General Charges.
Native Labour Commission.
Native Wages.
Chinese Wages.
Sundry Persons.
Forward
Drufontein
Angelo Interpreter
Newcomet 1st
Employees Room Guarantee A/c
Refunded the following deposits paid by them.
J. McNally £1 E. Sweetland £1
Boiler Licences A/c
Paid Licence on Boiler No 2657 New Comet Compound for the year ending 31st Dec 1906
Forward

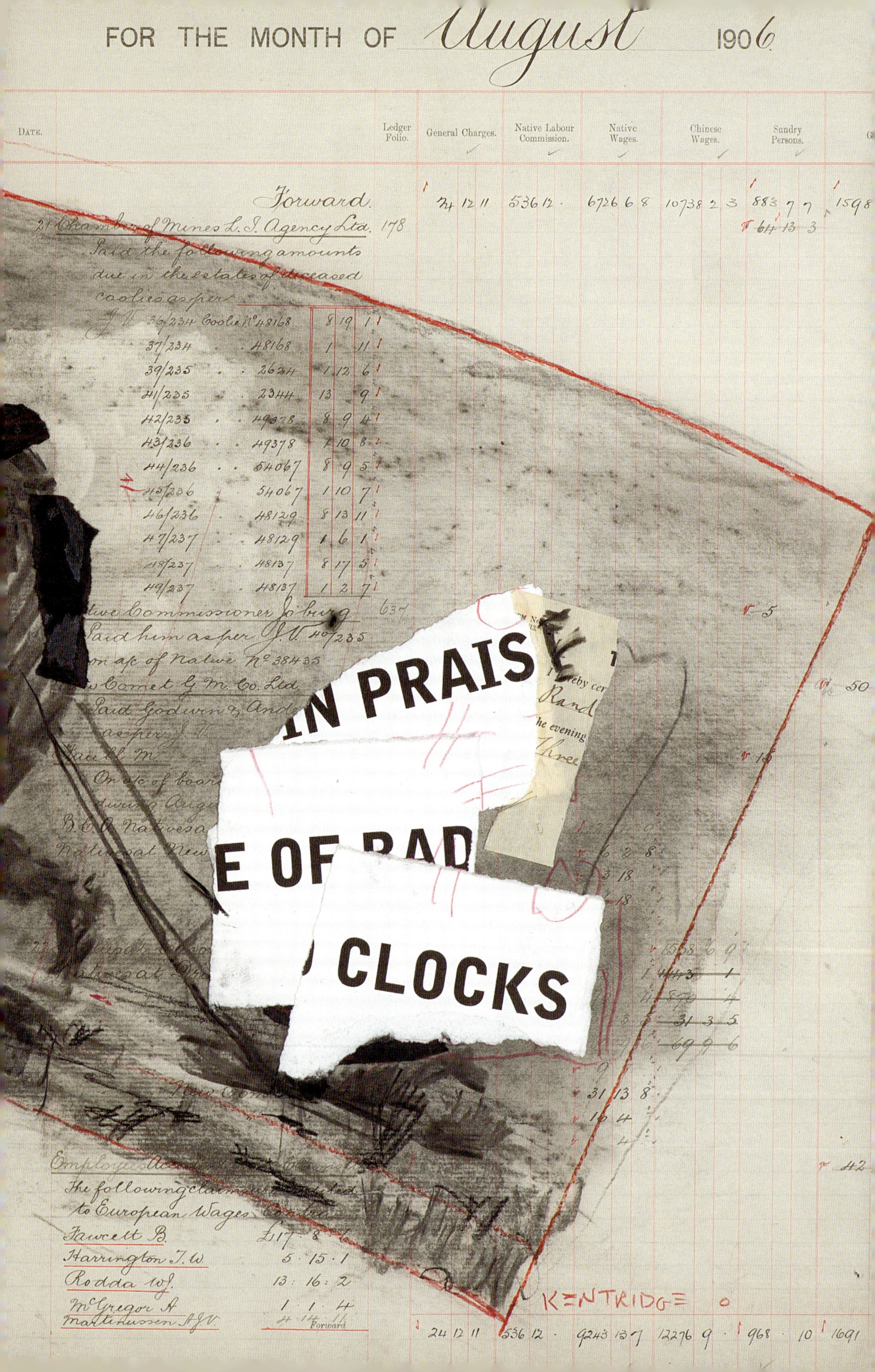

FOR THE MONTH OF August 1906
IN PRAIS
CLOCKS
KENTRIDGE

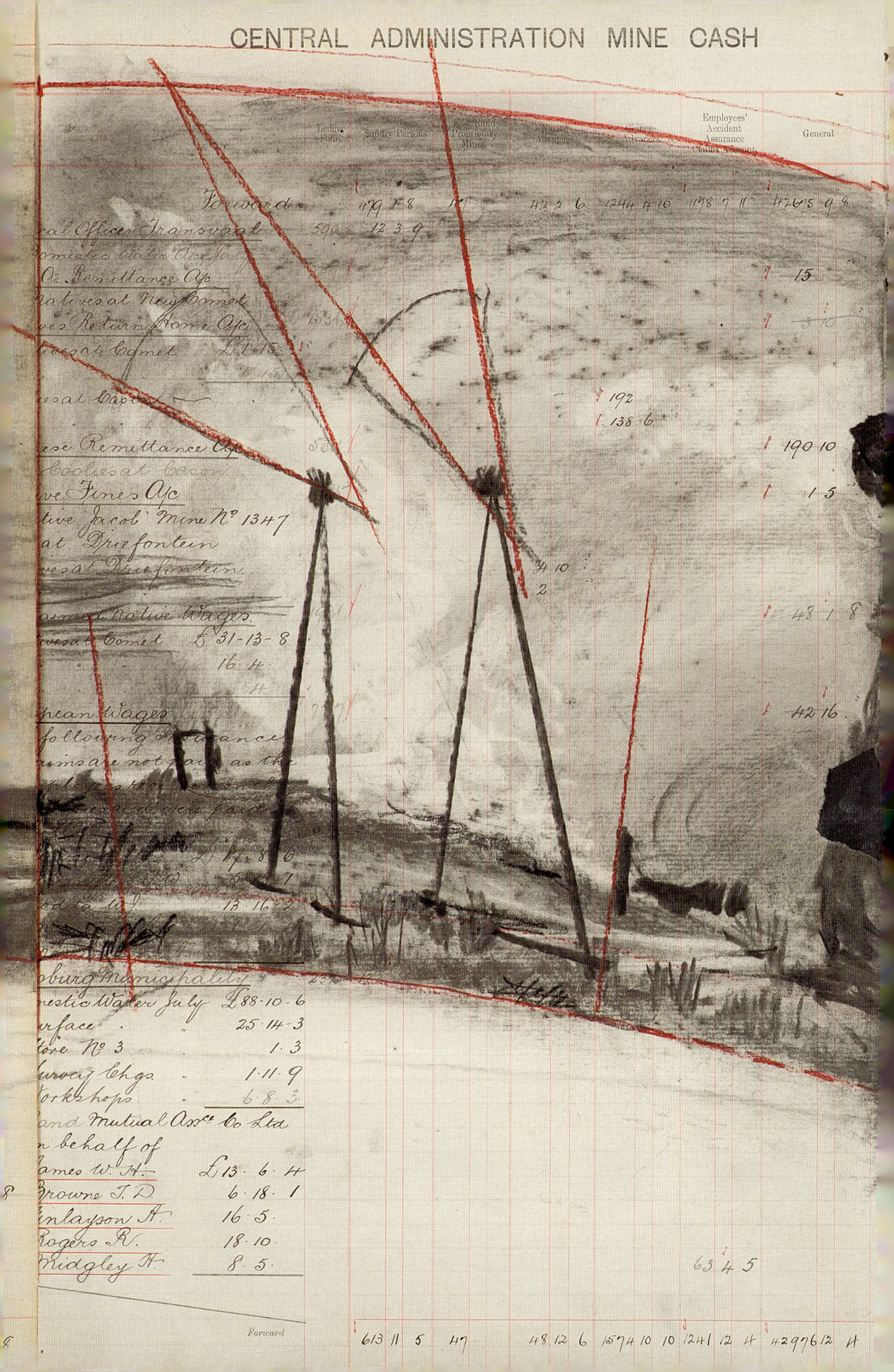
CENTRAL ADMINISTRATION MINE CASH
Ledger Folio
Sundry Persons
Employees' Accident Assurance
General
Forward
Forward

FOR THE MONTH OF ______________ 190

Date.		Ledger Folio.	General Charges.	Native Labour Commission.	Native Wages.	Chinese Wages.	Sundry Persons.	General.
	Forward							

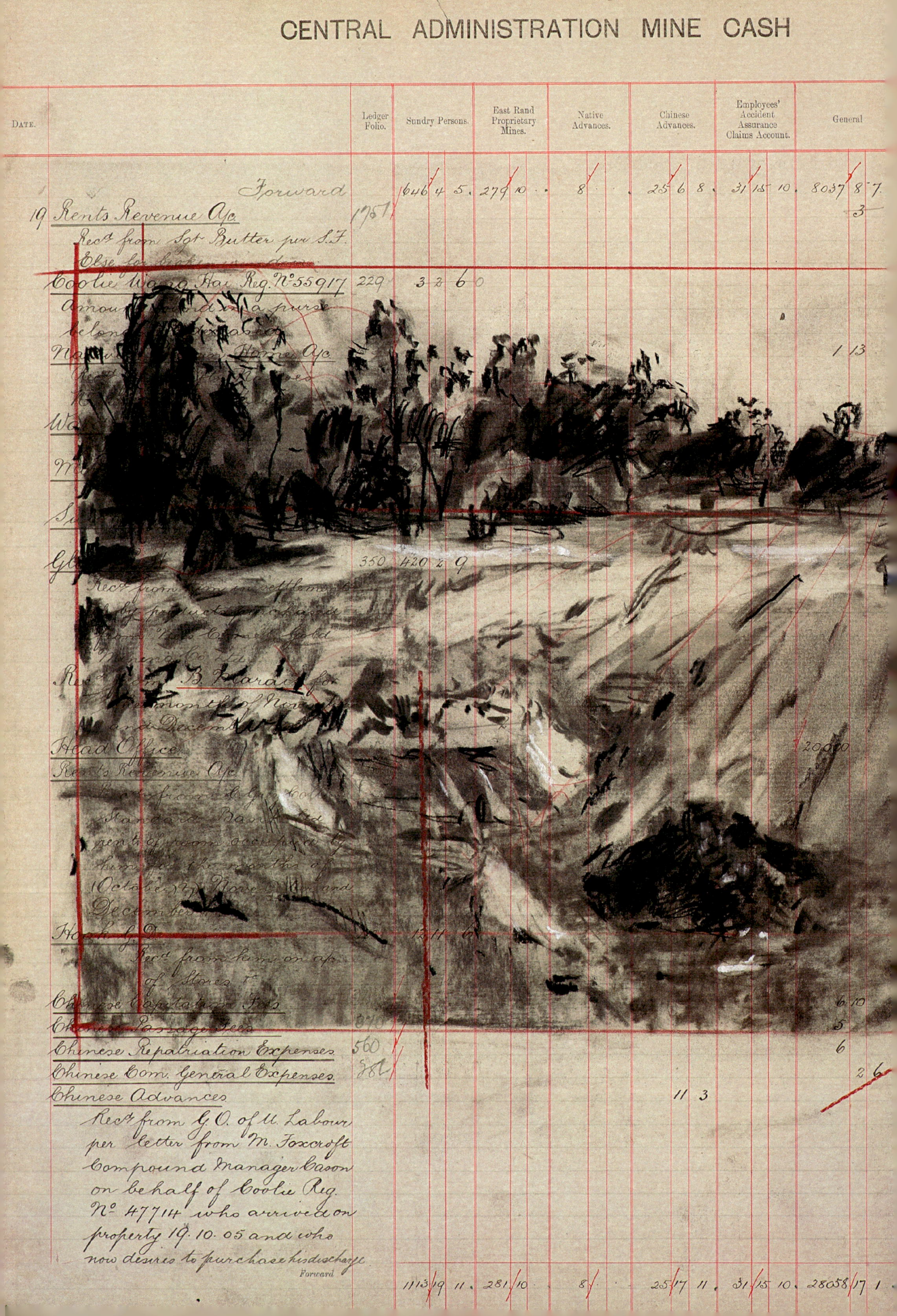

CENTRAL ADMINISTRATION MINE CASH
DATE.
Ledger Folio.
Sundry Persons.
East Rand Proprietary Mines.
Native Advances.
Chinese Advances.
Employees' Accident Assurance Claims Account.
General
Forward
Rents Revenue A/c
Coolie Wong Hai Reg. No 55917
Head Office
Chinese Repatriation Expenses
Chinese Com. General Expenses
Chinese Advances
Rec'd from G.O. of N. Labour per letter from M. Foxcroft Compound Manager Cason on behalf of Coolie Reg. No 47714 who arrived on property 19.10.05 and who now desires to purchase his discharge
Forward

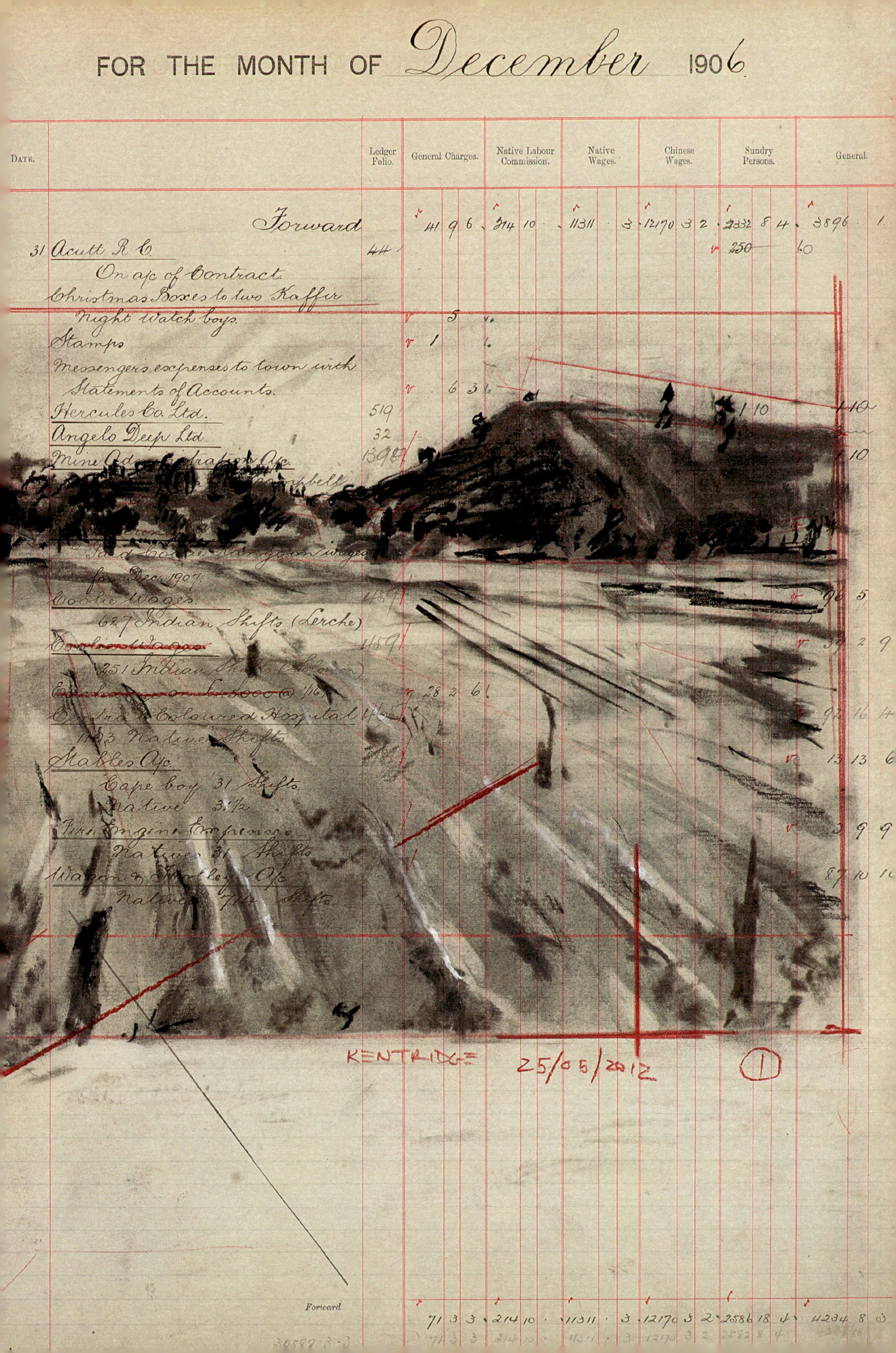
FOR THE MONTH OF December 1906
Date.
Ledger Folio.
General Charges.
Native Labour Commission.
Native Wages.
Chinese Wages.
Sundry Persons.
General.
Forward
31 Acutt R. C.
On a/c of Contract.
Christmas Boxes to two Kaffir
Night Watch boys.
Stamps
Messengers expenses to town with
Statements of Accounts.
Hercules Co. Ltd.
Angelo Deep Ltd.
Coolie Wages
627 Indian Shifts (Lerche)
Stables A/c.
Cape boy 31 Shifts.
Native 31 1/2
Forward
KENTRIDGE 25/05/2012

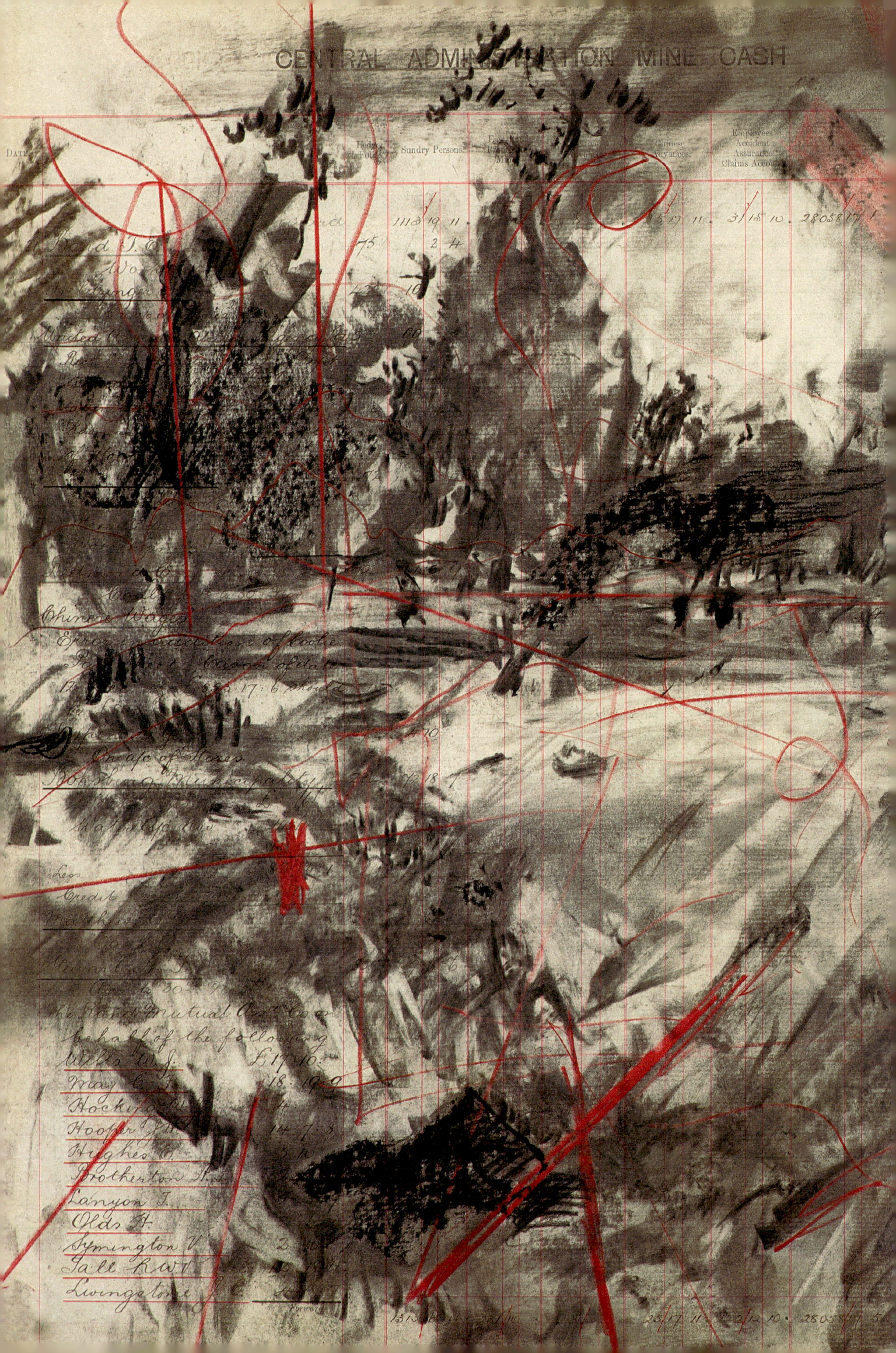

CENTRAL ADMINISTRATION MINE CASH
Sundry Persons
Lanyon J.
Olds H.
Symington V
Livingstone

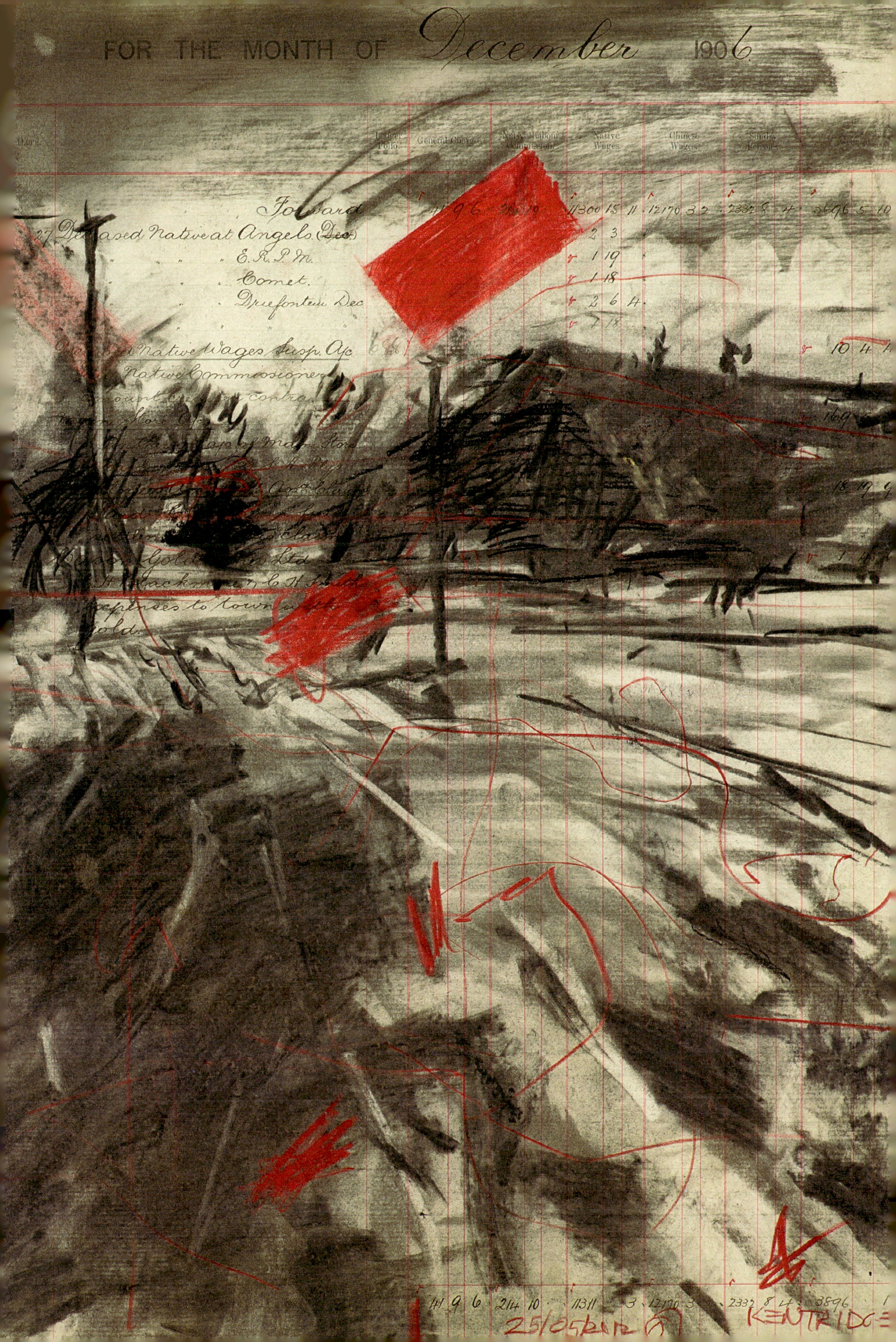
FOR THE MONTH OF December 1906
Forward
Deceased Native at Angels
E.R.P.M.
Comet.
Driefontein Dee
Native Wages Susp. A/c
Native Commissioner
KENTRIDGE

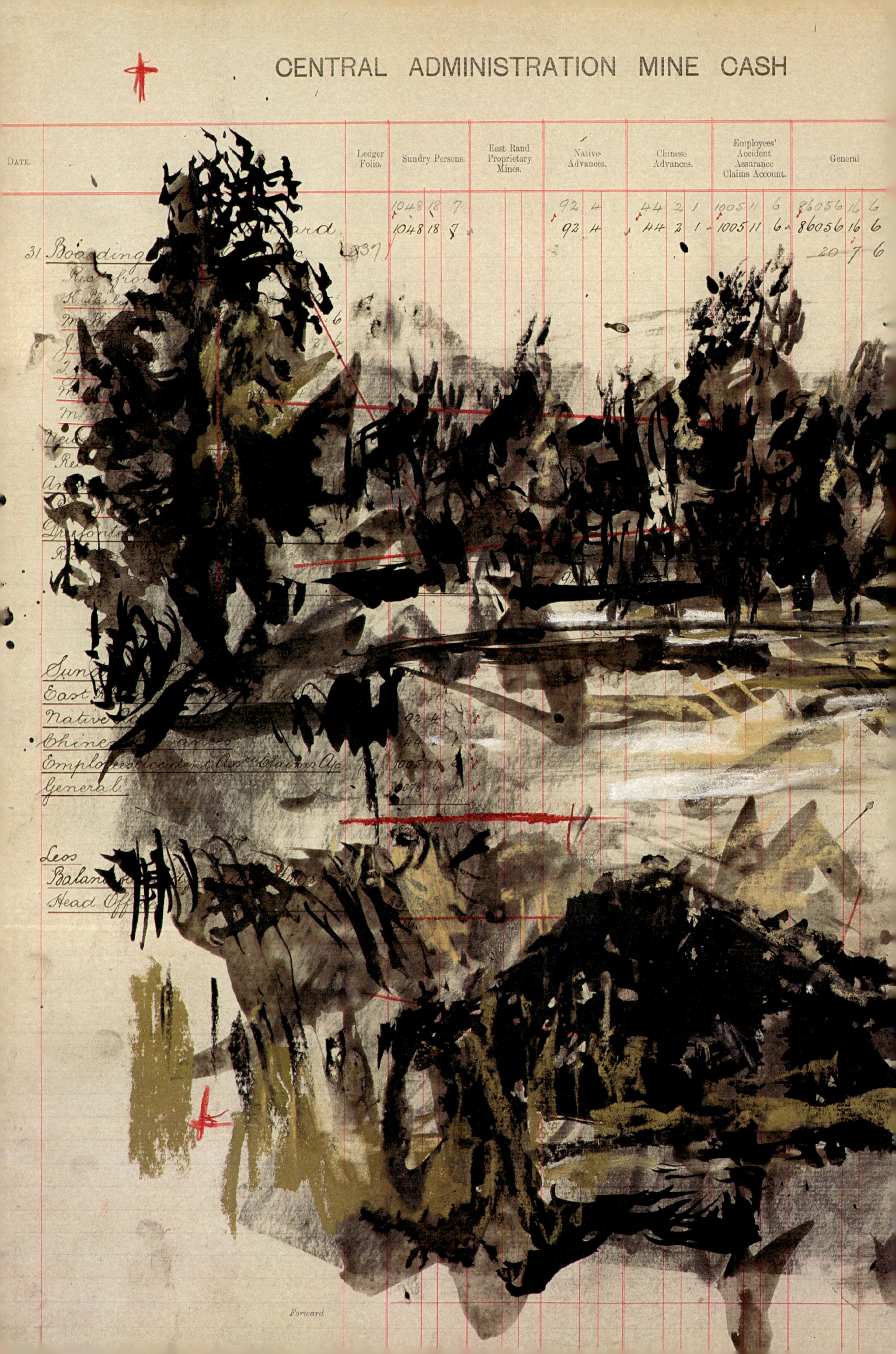
CENTRAL ADMINISTRATION MINE CASH
DATE.
Ledger Folio.
Sundry Persons.
East Rand Proprietary Mines.
Native Advances.
Chinese Advances.
Employees' Accident Assurance Claims Account.
General
Forward

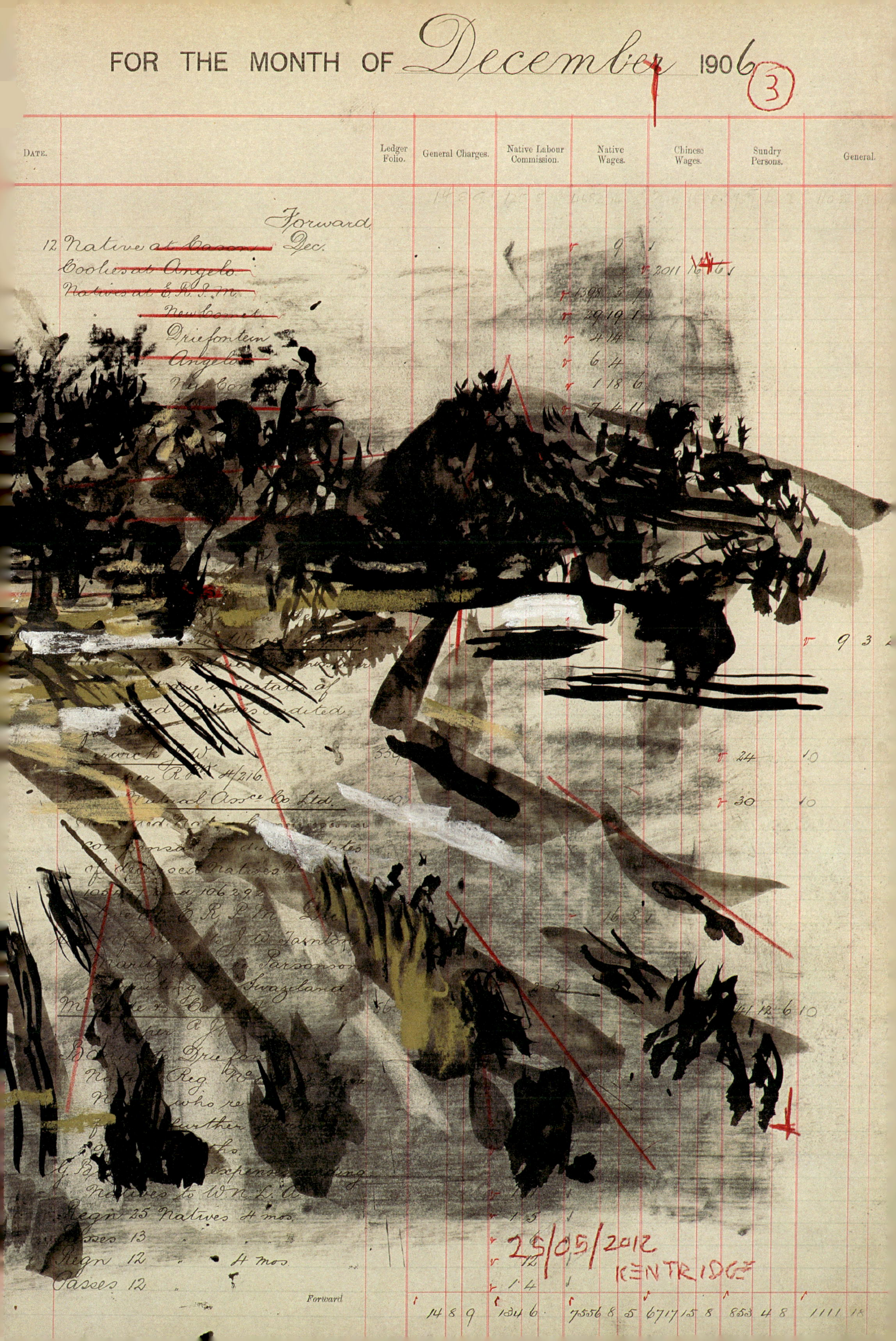

FOR THE MONTH OF December 1906
3
Date.
Ledger Folio.
General Charges.
Native Labour Commission.
Native Wages.
Chinese Wages.
Sundry Persons.
General.
Forward Dec.
Forward
25/05/2012
KENTRIDGE

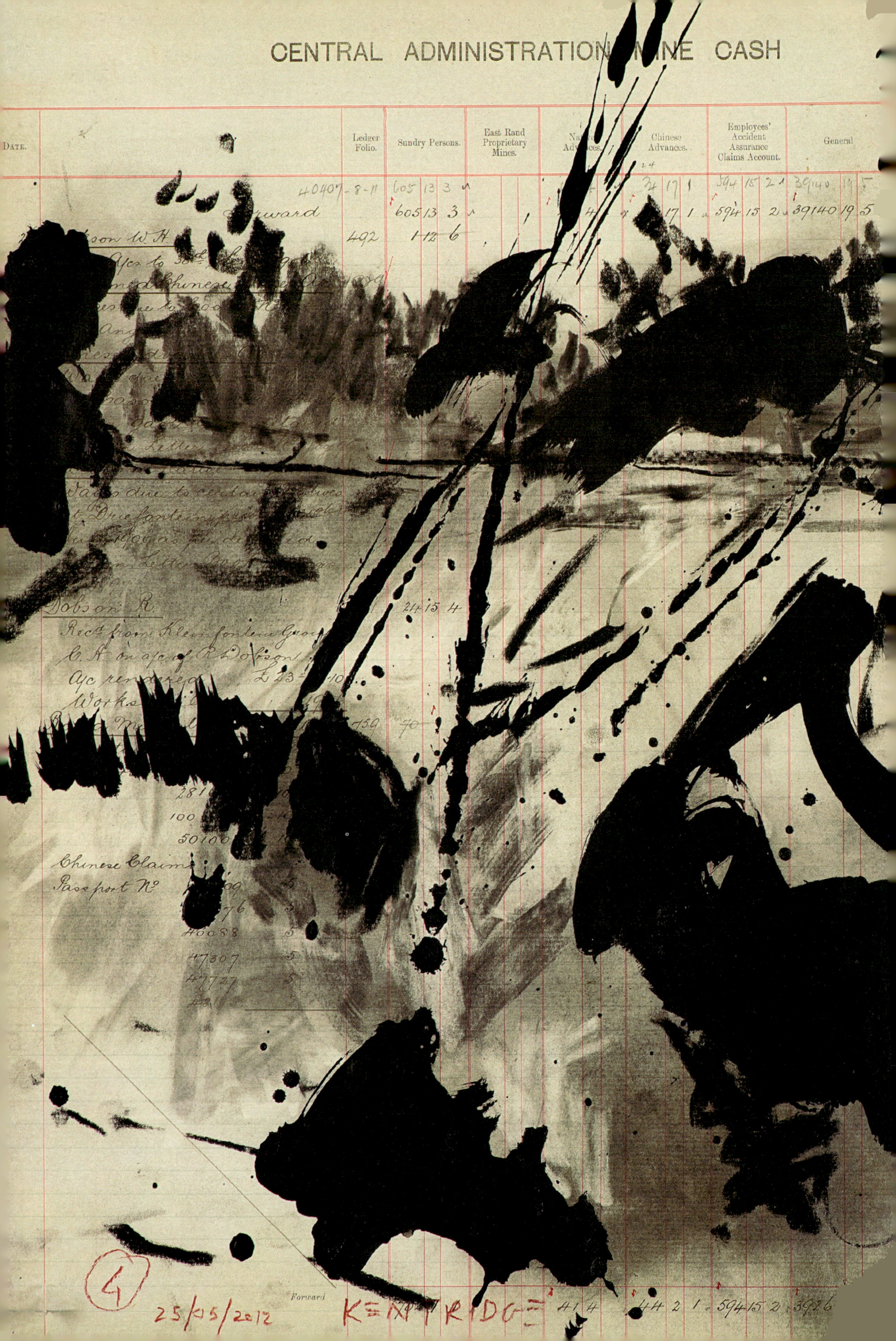
CENTRAL ADMINISTRATION MINE CASH
Date.
Ledger Folio.
Sundry Persons.
East Rand Proprietary Mines.
Chinese Advances.
Employees' Accident Assurance Claims Account.
General
Dobson R.
Chinese Claims
Passport No
Forward
4
25/05/2012
KENTRIDGE

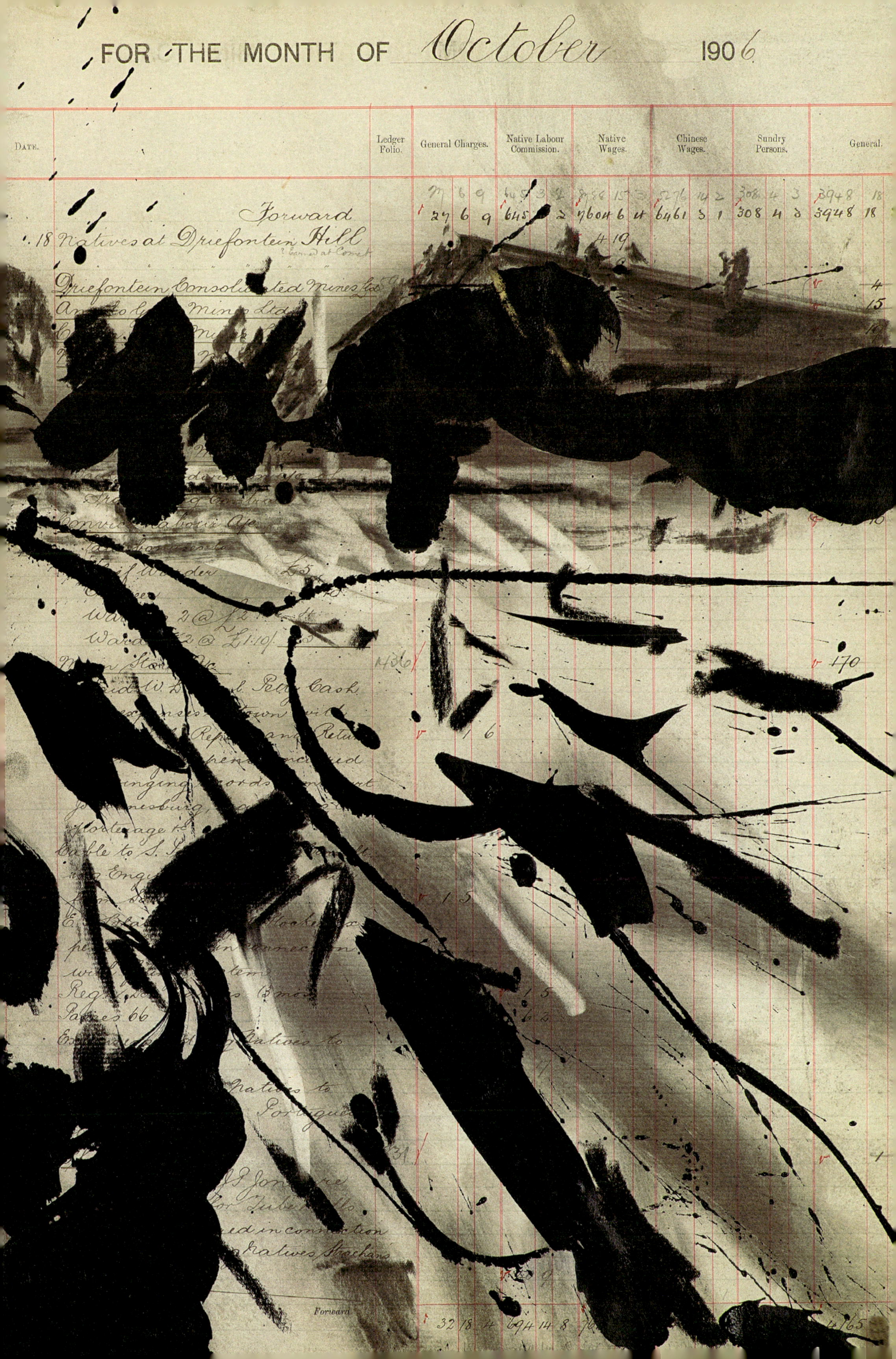
FOR THE MONTH OF October 1906
Date.
Ledger Folio.
General Charges.
Native Labour Commission.
Native Wages.
Chinese Wages.
Sundry Persons.
General.
Forward
18 Natives at Driefontein Hill
Driefontein Consolidated Mines Ltd
Forward

CENTRAL ADMINISTRATION MINE CASH
DATE.
Ledger Folio.
Sundry Persons.
Native Advances.
Employees' Accident Assurance Claims Account.
General.
Forward

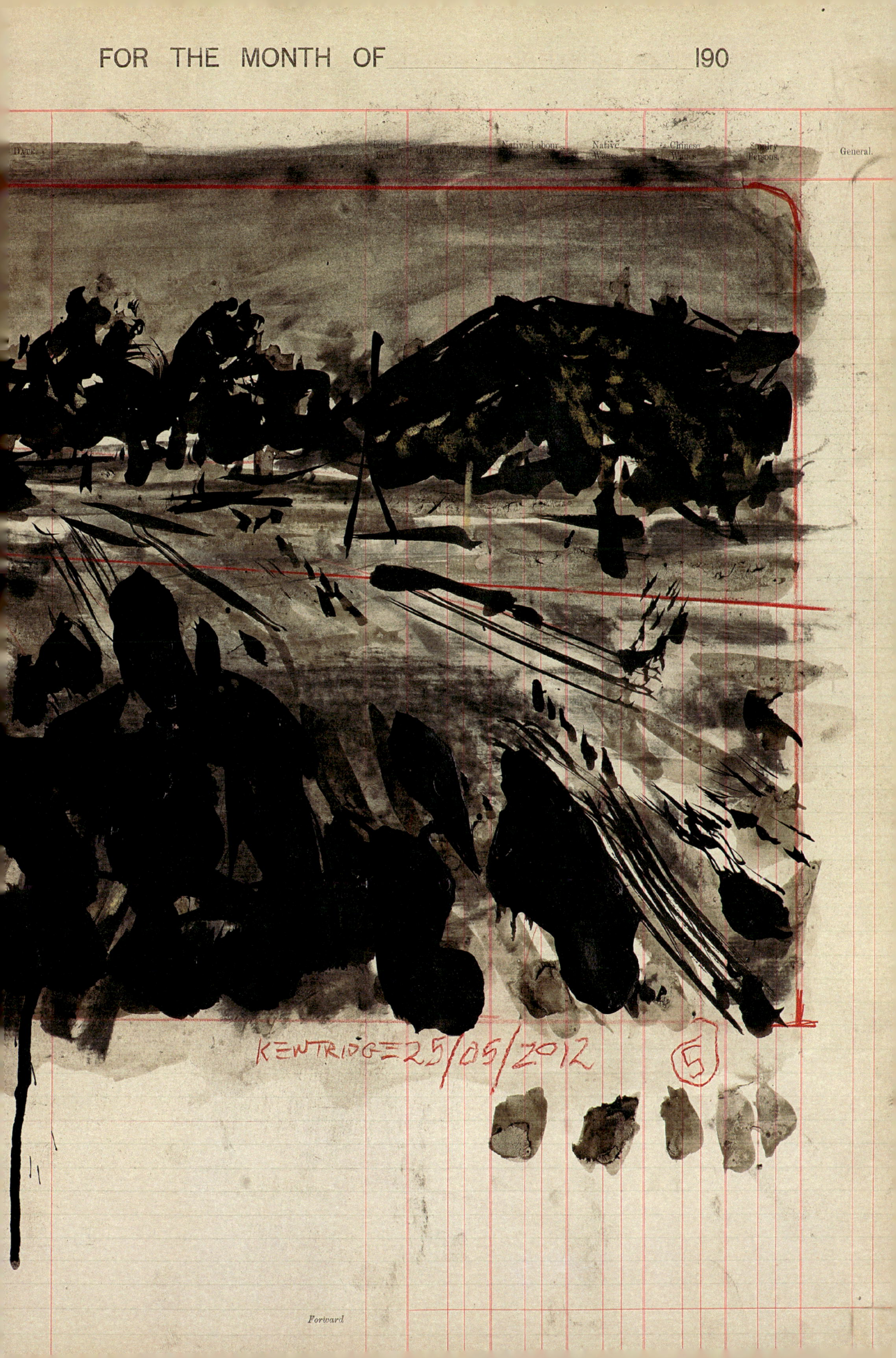
FOR THE MONTH OF 190
General.
KENTRIDGE 25/05/2012
5
Forward

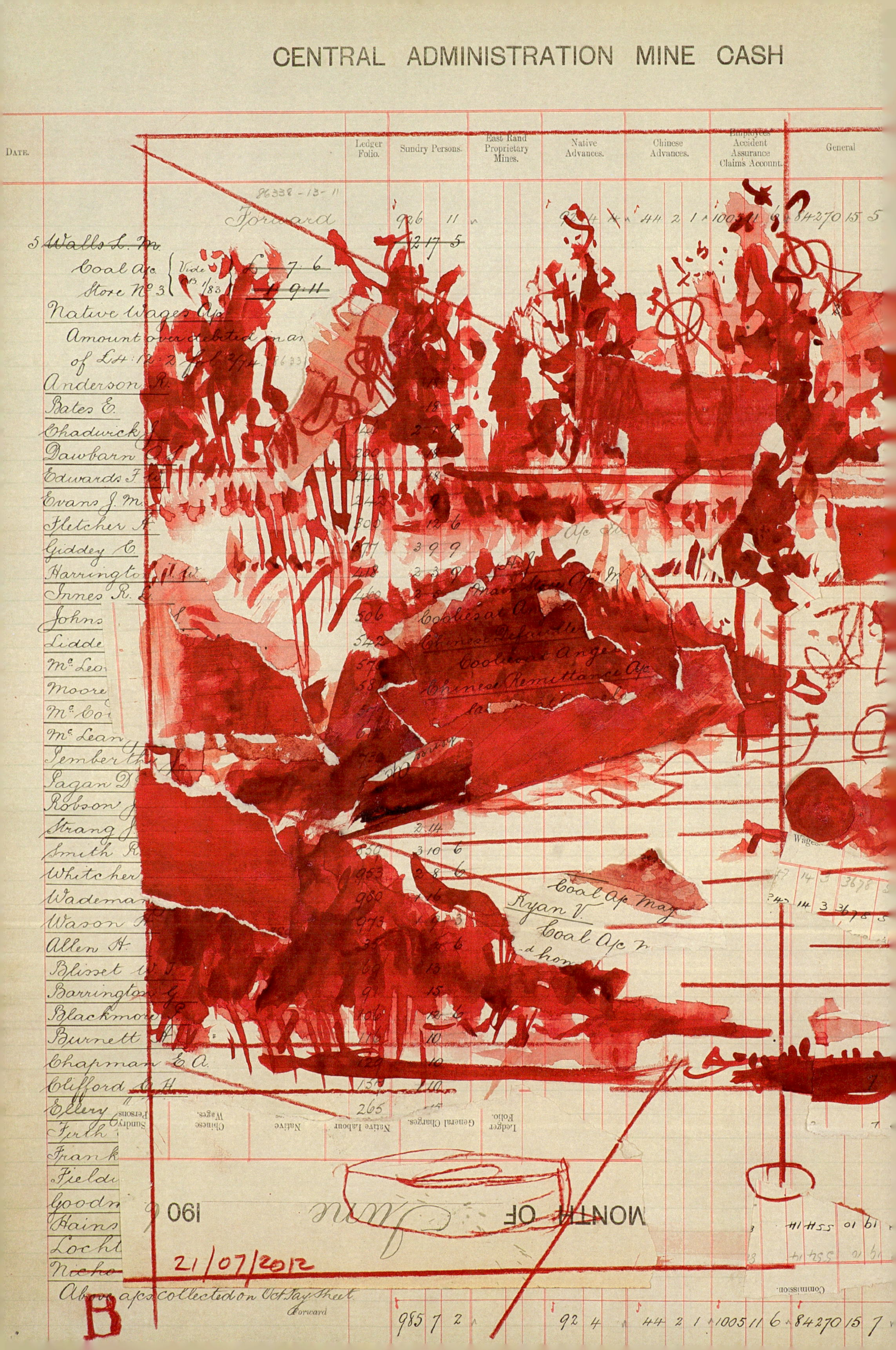
CENTRAL ADMINISTRATION MINE CASH
Date.
Ledger Folio.
Sundry Persons.
East Rand Proprietary Mines.
Native Advances.
Chinese Advances.
Employees Accident Assurance Claims Account.
General
Forward
Coal A/c
Store No 3
Native Wages A/c
Anderson R.
Bates E.
Chadwick
Dawbarn
Edwards F.
Evans J. M.
Fletcher
Giddey C.
Innes R.
Allen H.
Chapman E.A.
Clifford
Coolies at Angelo
Chinese Remittance A/c
Coal A/c
Ryan V.
MONTH OF
190
21/07/2012
B
Commission.

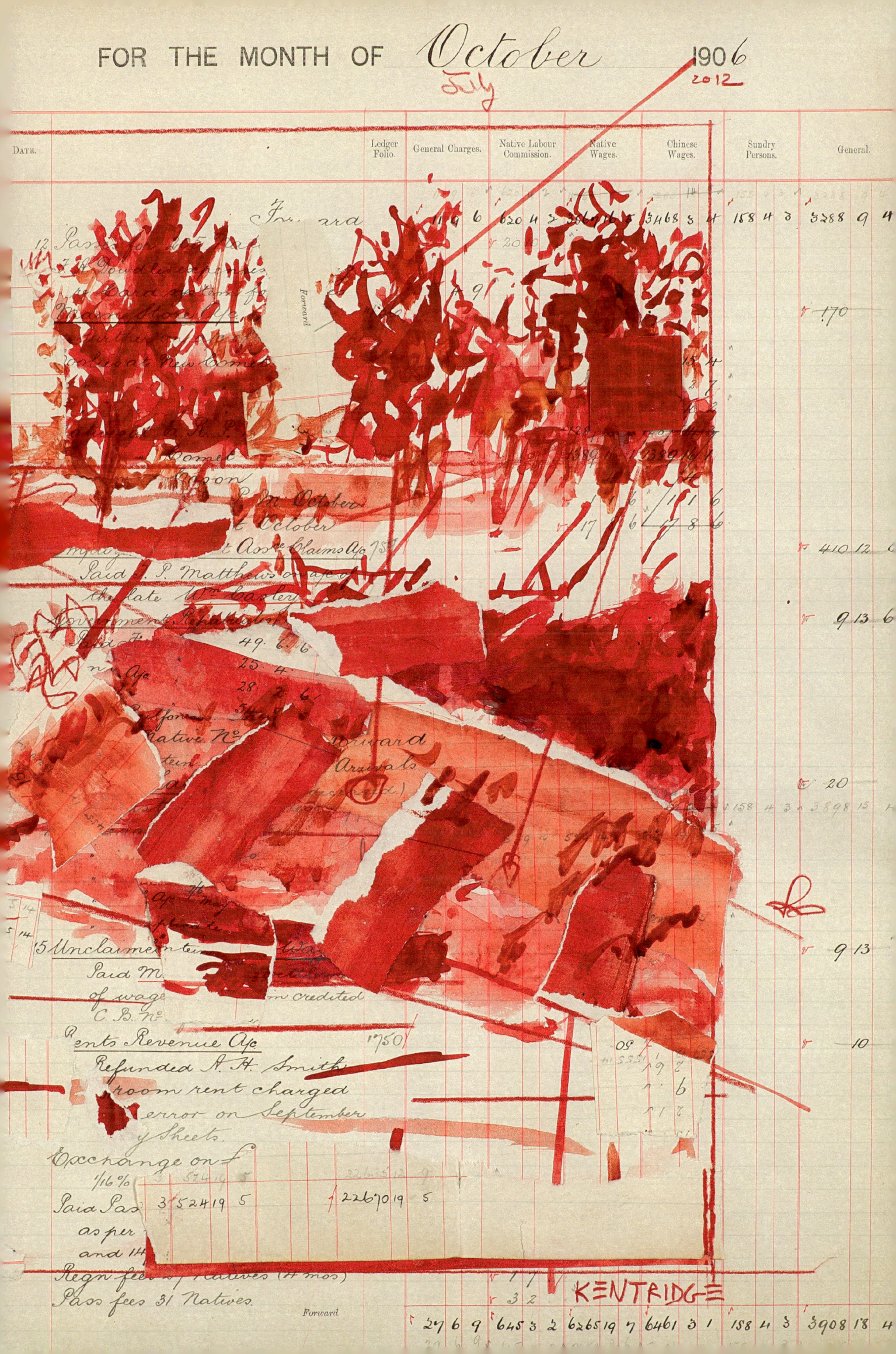
FOR THE MONTH OF October 1906
July 2012
Date.
Ledger Folio.
General Charges.
Native Labour Commission.
Native Wages.
Chinese Wages.
Sundry Persons.
General.
Forward
Rents Revenue A/c
Refunded A. H. Smith
room rent charged
error on September
Exchange on £
Pass fees 31 Natives.
Forward
KENTRIDGE

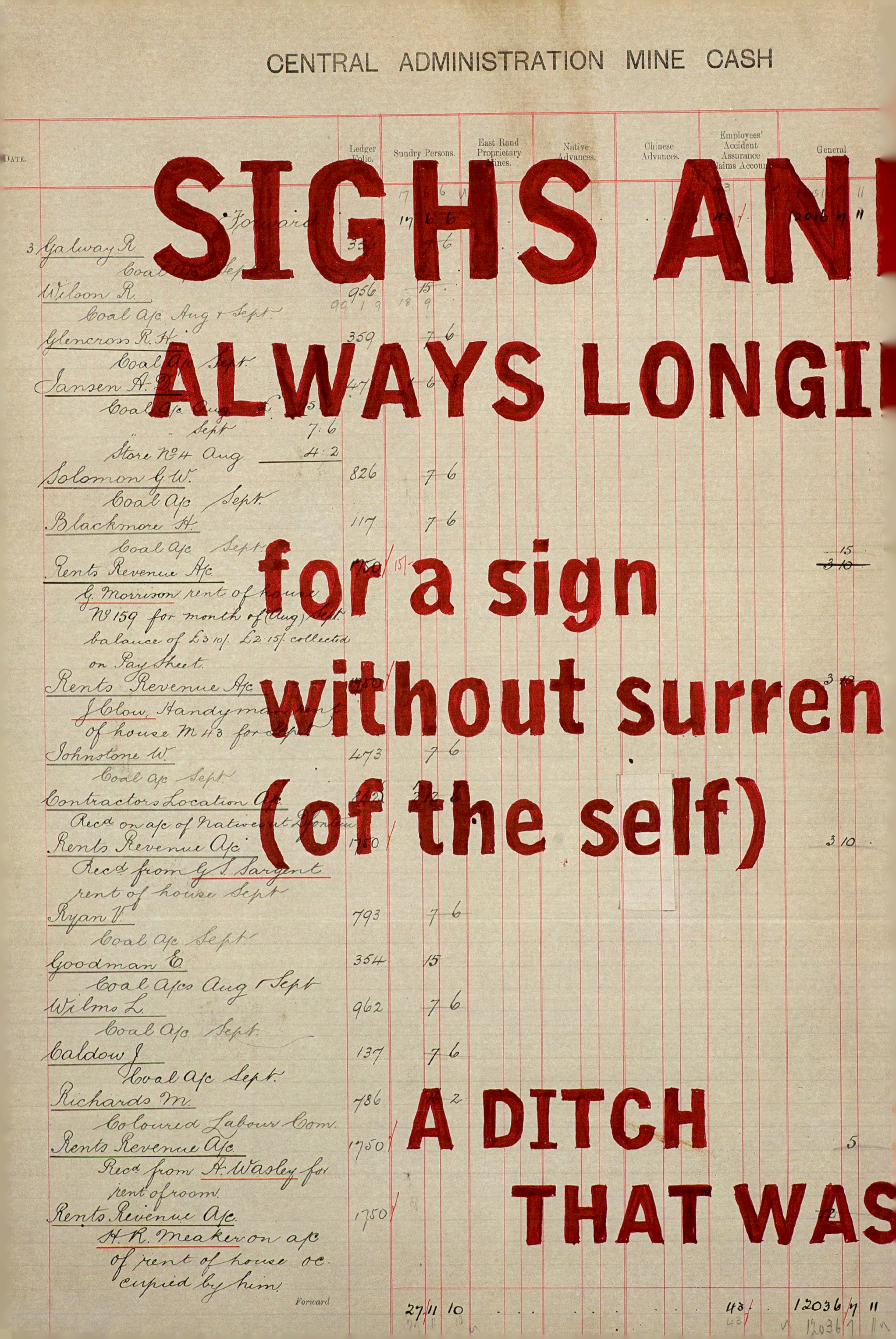
CENTRAL ADMINISTRATION MINE CASH
SIGHS AN
ALWAYS LONGI
for a sign
without surren
(of the self)
A DITCH
THAT WAS

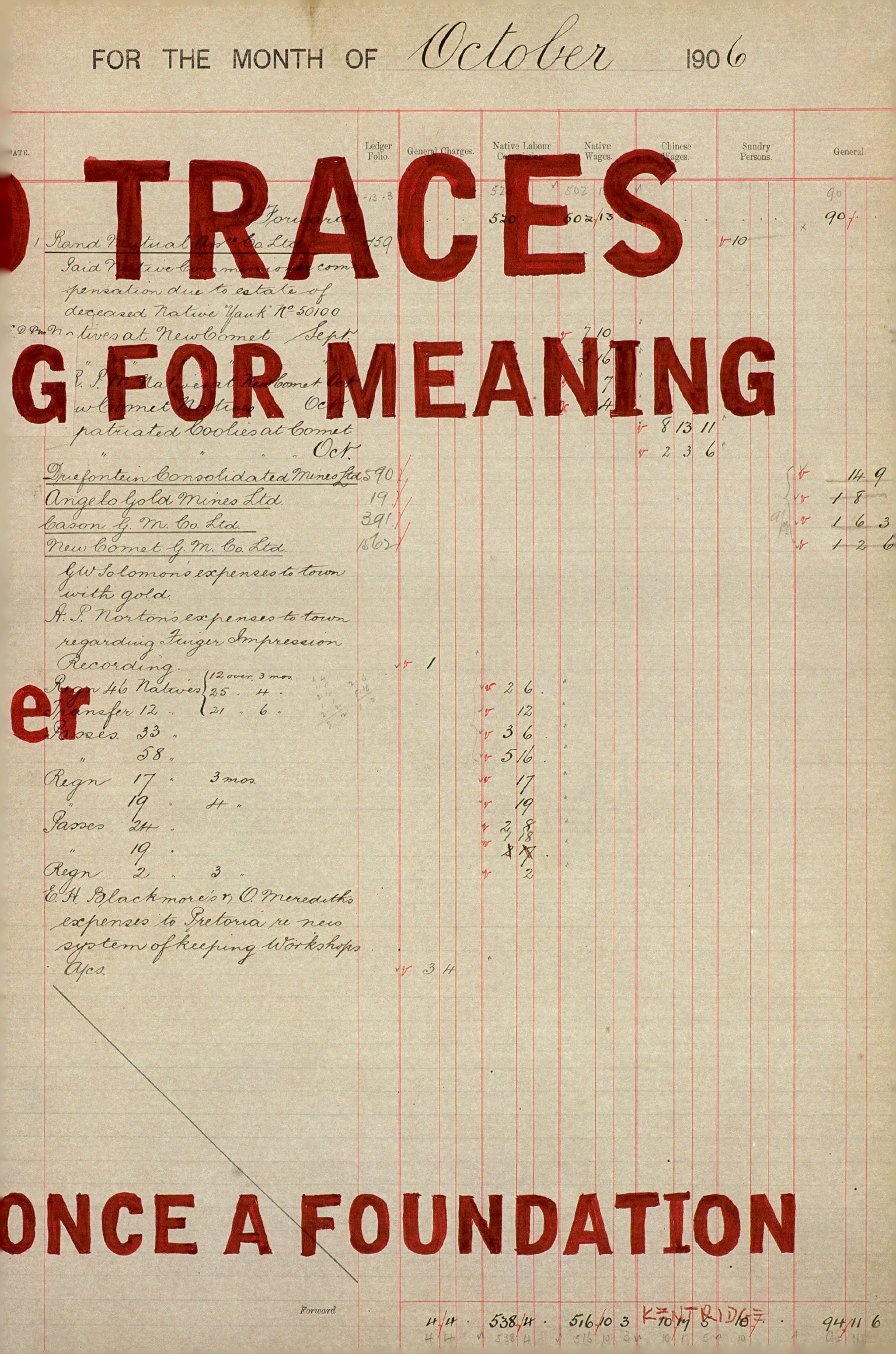
FOR THE MONTH OF October 1906
Ledger Folio.
General Charges.
Native Labour Commission.
Native Wages.
Chinese Wages.
Sundry Persons.
General.
TRACES
G FOR MEANING
er
ONCE A FOUNDATION
Forward
KENTRIDGE

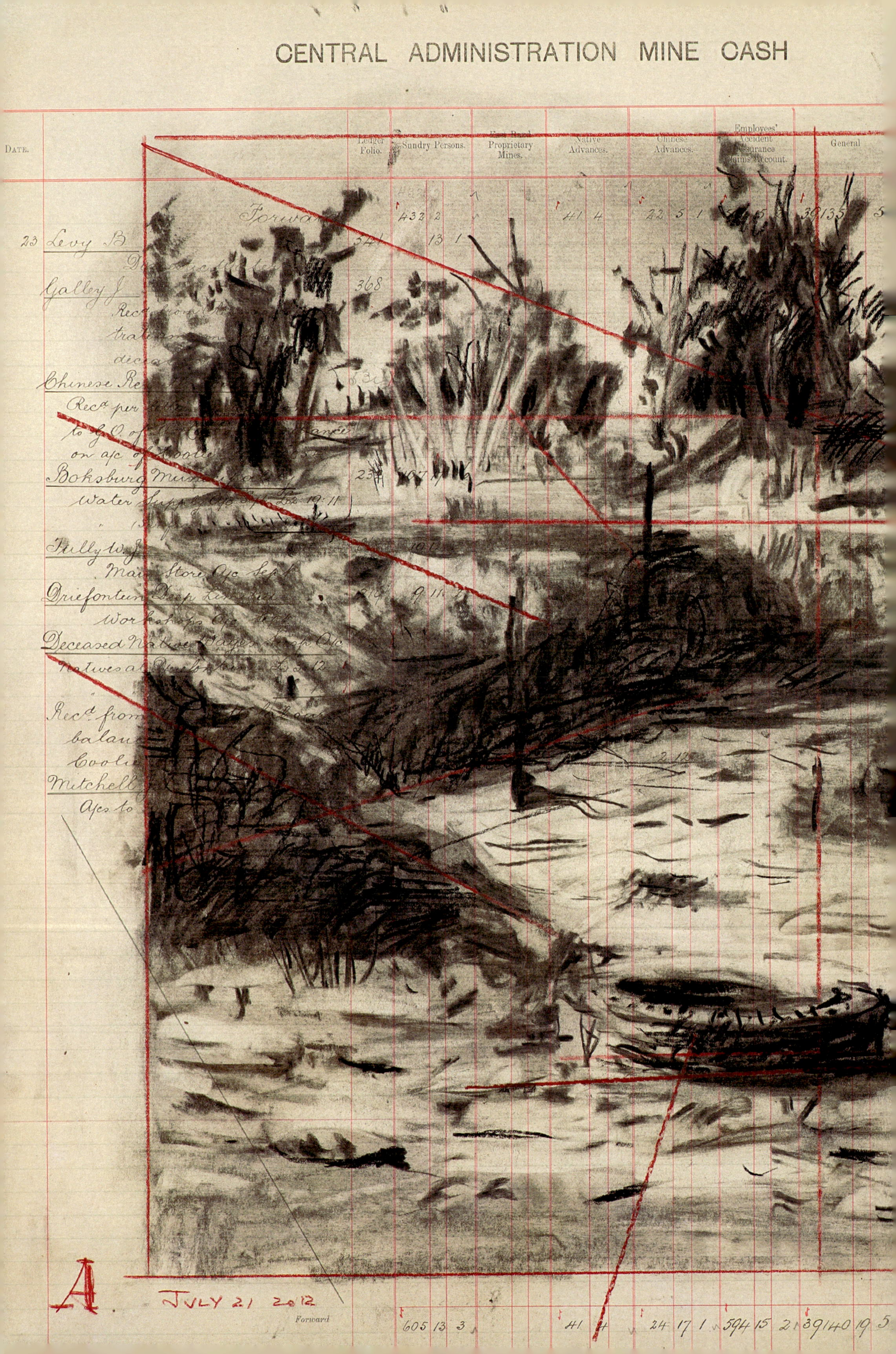

CENTRAL ADMINISTRATION MINE CASH
Date.
Ledger Folio.
Sundry Persons.
Proprietary Mines.
Native Advances.
Chinese Advances.
General
Levy B
Galley J
Boksburg Muni
Driefontein Deep
Deceased Native Wages
Forward
JULY 21 2012

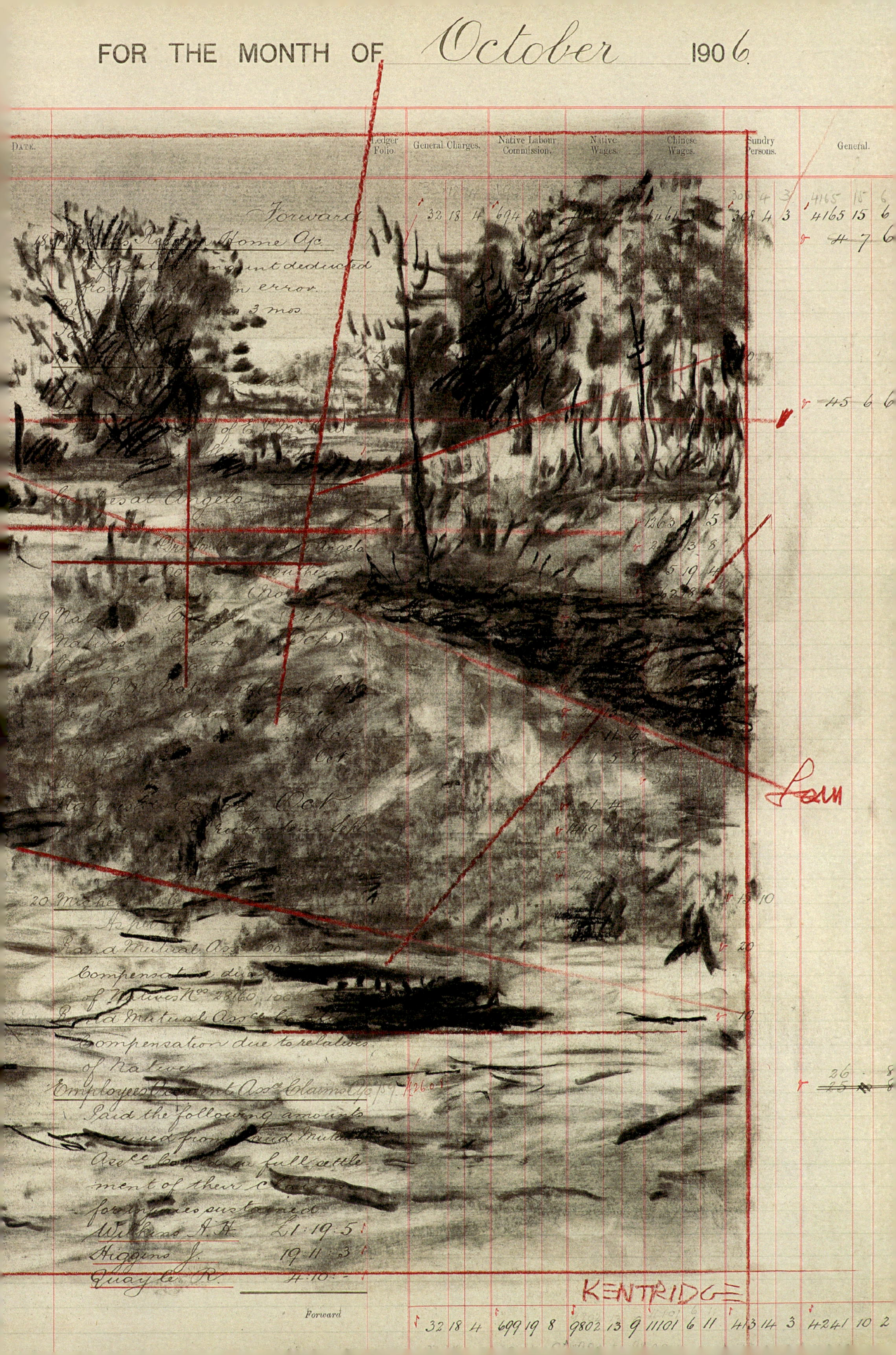
FOR THE MONTH OF October 1906
Date.
Ledger Folio.
General Charges.
Native Labour Commission.
Native Wages.
Chinese Wages.
Sundry Persons.
General.
Forward
Compensation due to relatives
of Native
Paid the following amounts
Wilkins A.H £1 19 5
Higgins J. 19 11 3
Quayle R. 4 10 –
Forward
KENTRIDGE

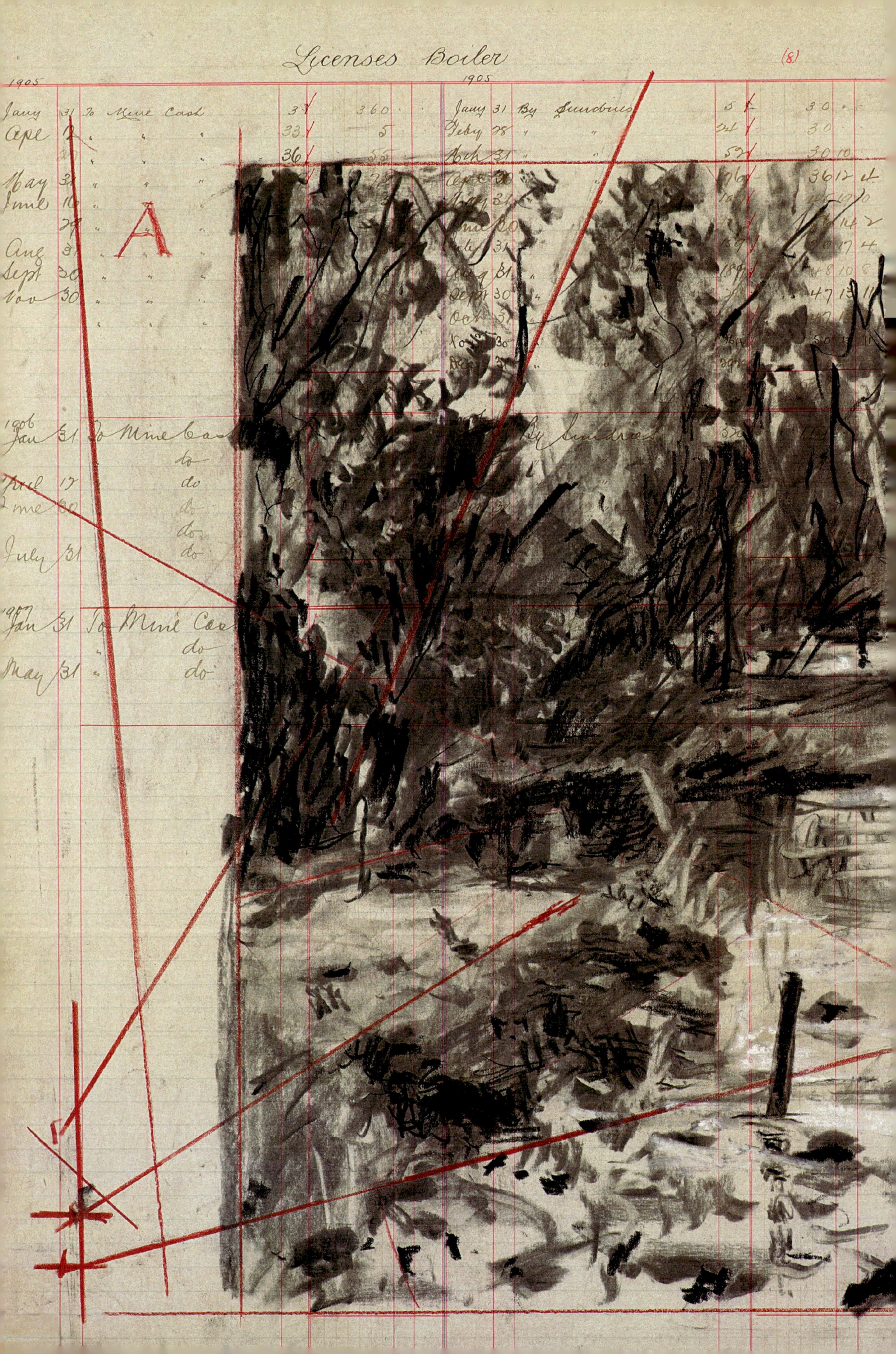
Licenses Boiler
(8)
1905
A

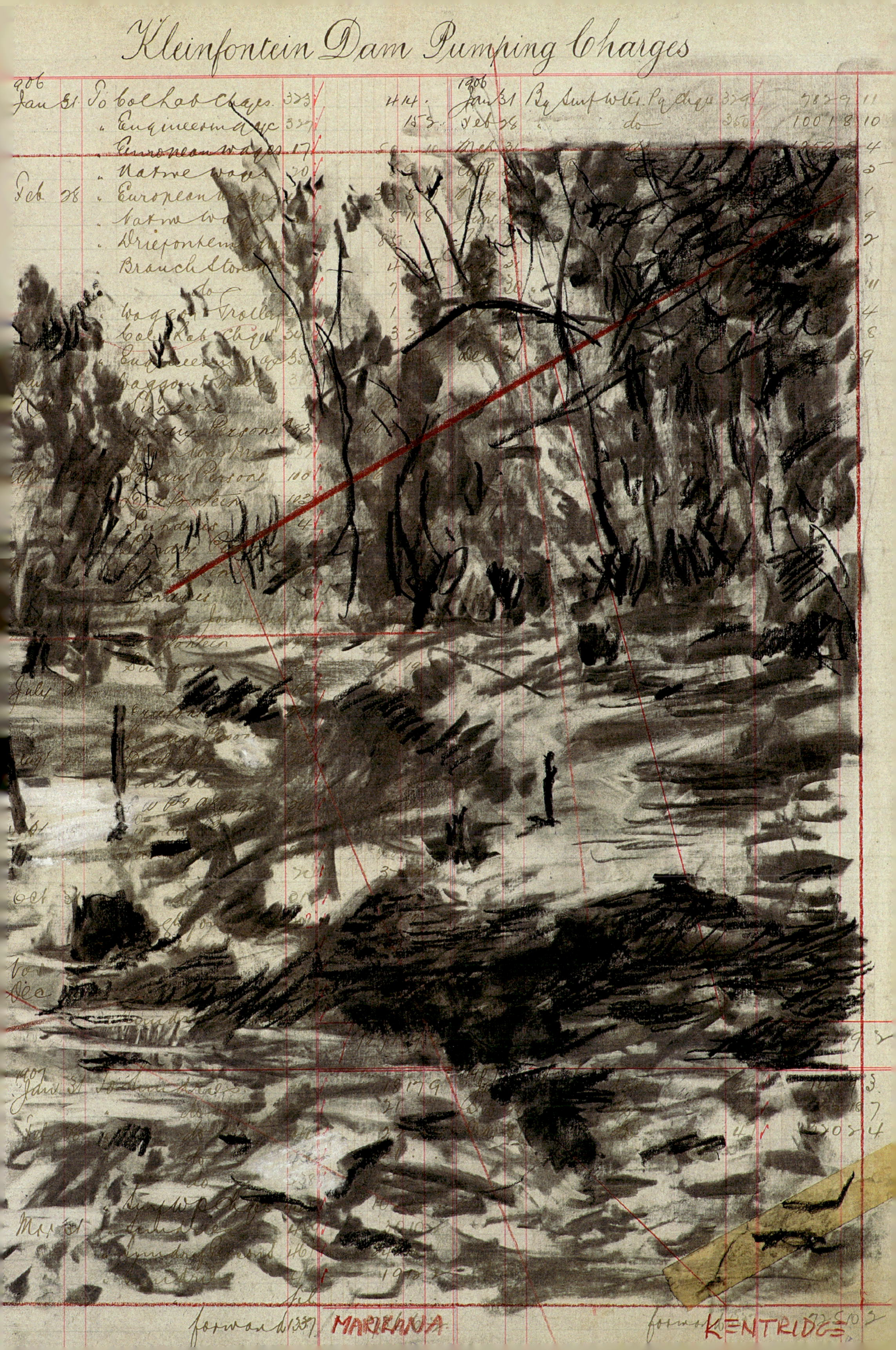
Kleinfontein Dam Pumping Charges
MARIKANA
KENTRIDGE

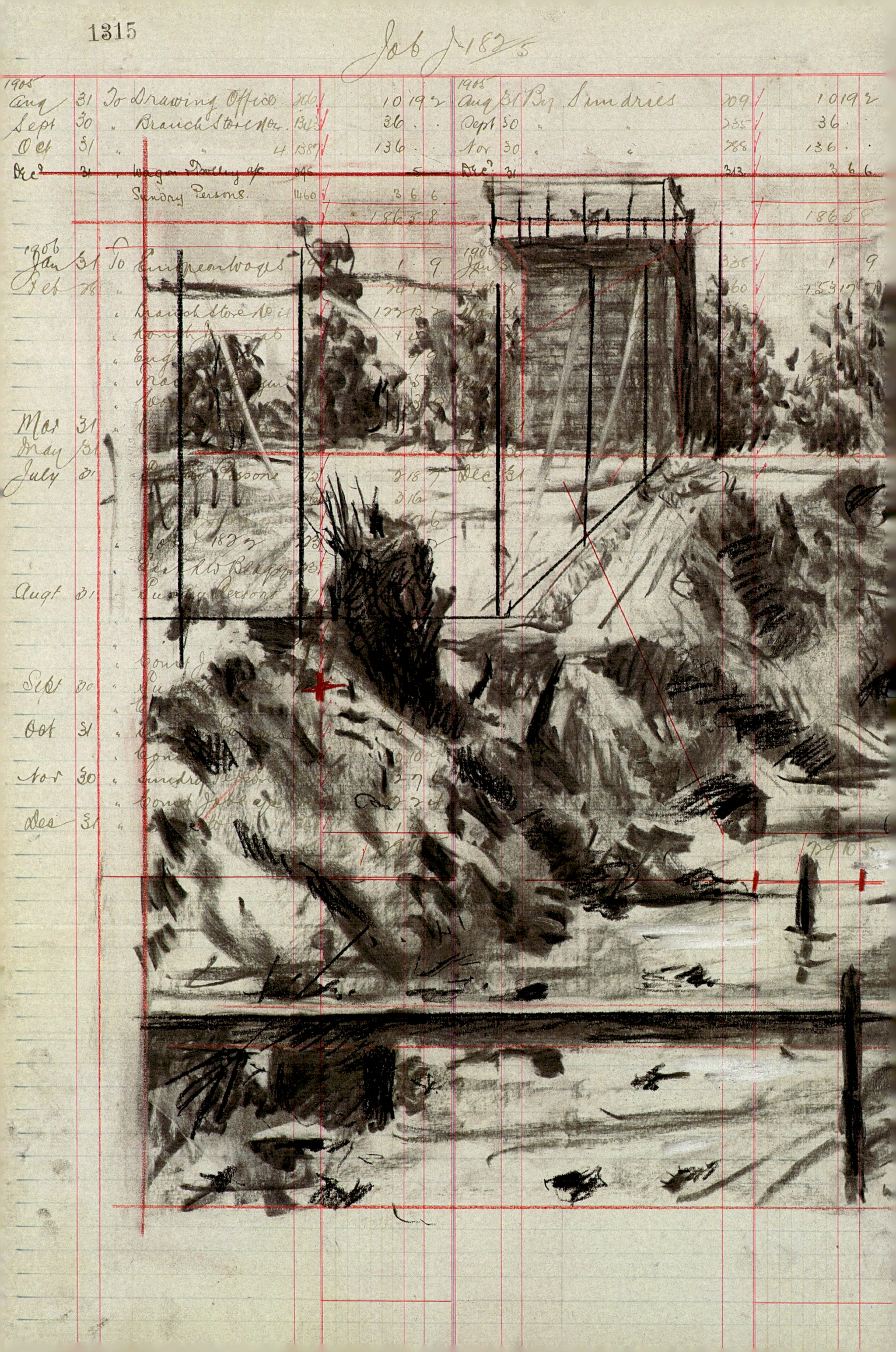
1315
To Drawing Office
By Sundries

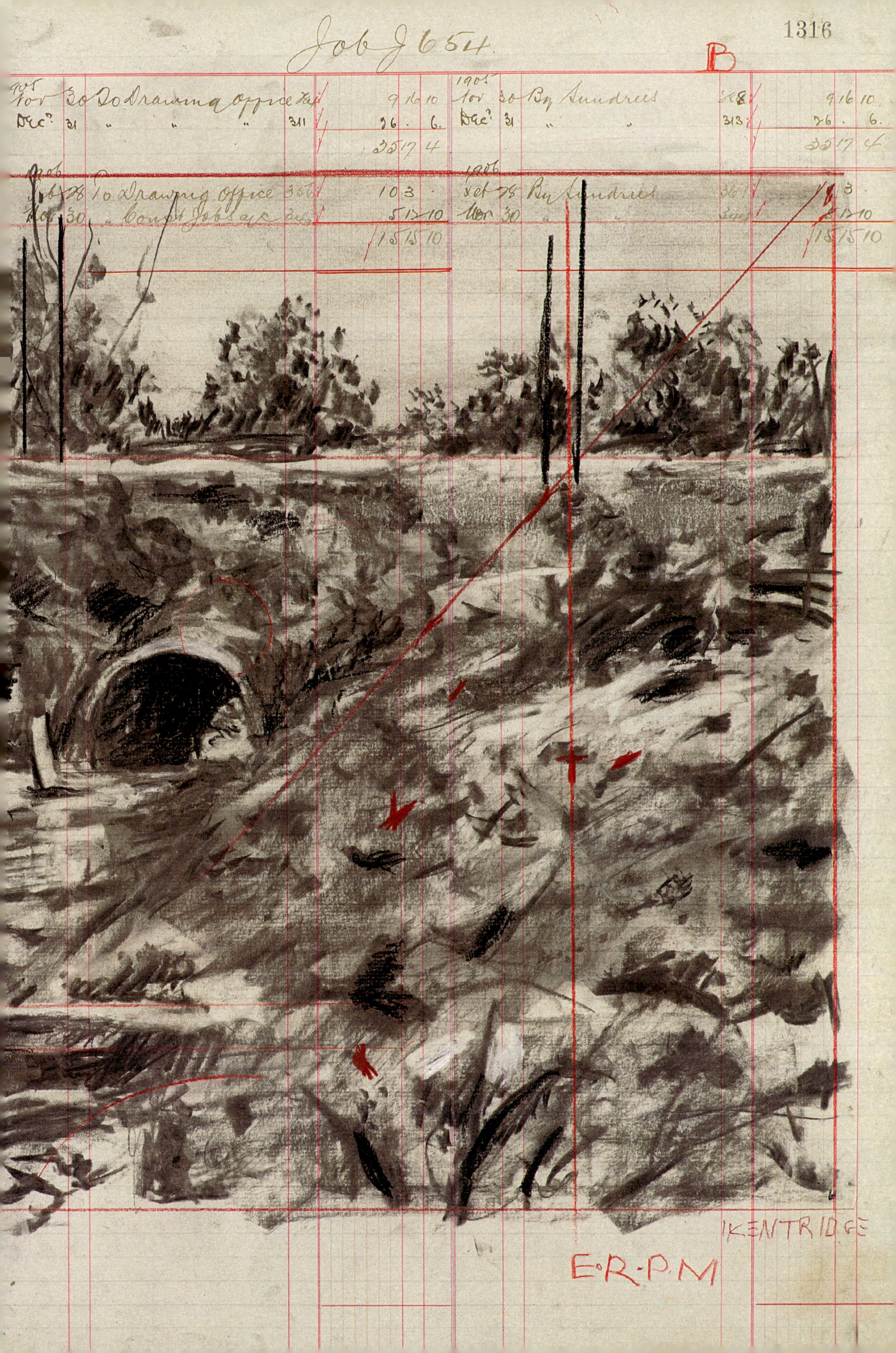
1316
Job J 654
B
To Drawing Office
By Sundries
E·R·P·M
KENTRIDGE

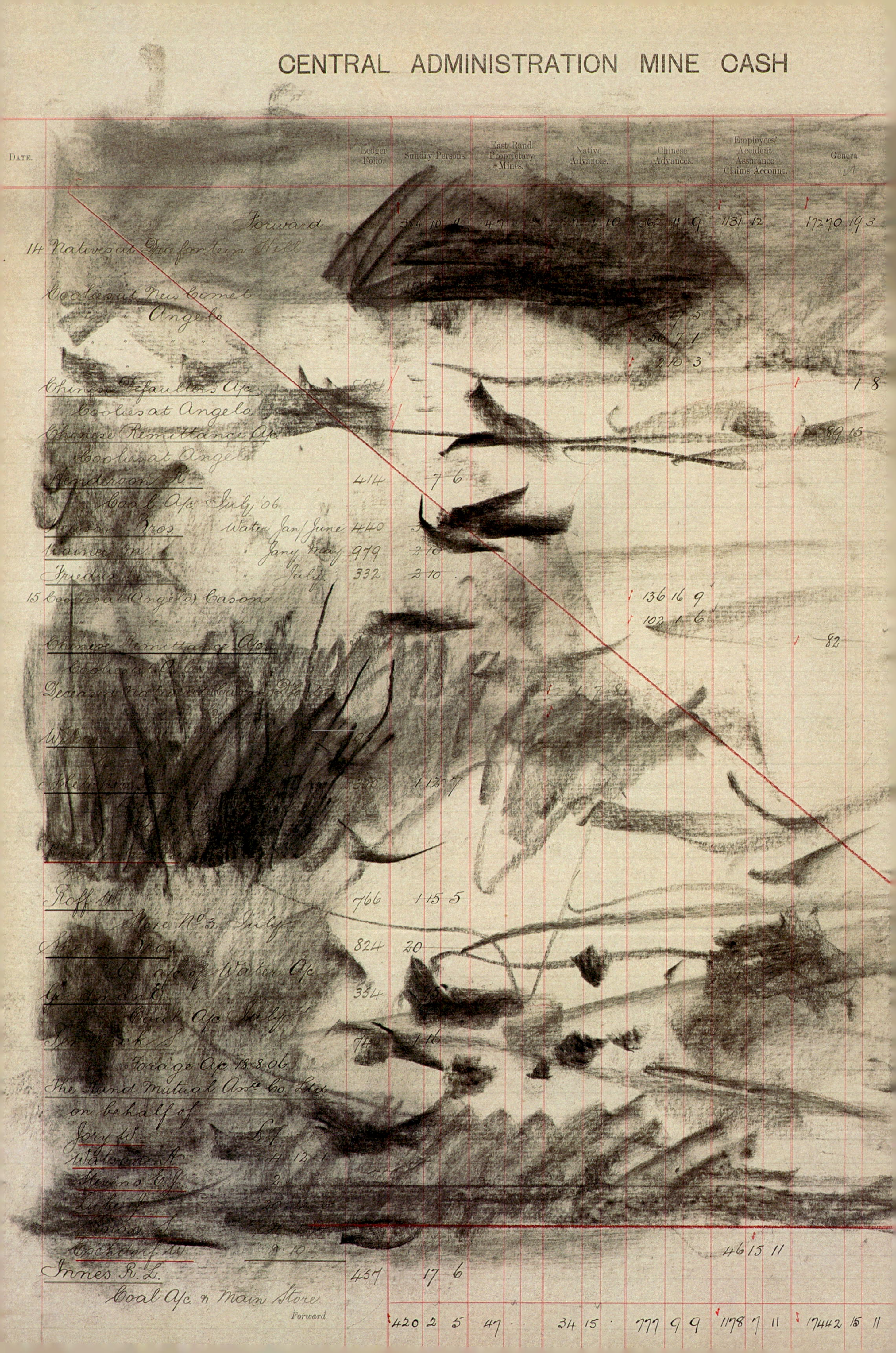
CENTRAL ADMINISTRATION MINE CASH
Date.
Ledger Folio
Sundry Persons
East Rand Proprietary Mines
Native Advances
Chinese Advances
Employees' Accident Assurance Claims Account
General
Forward

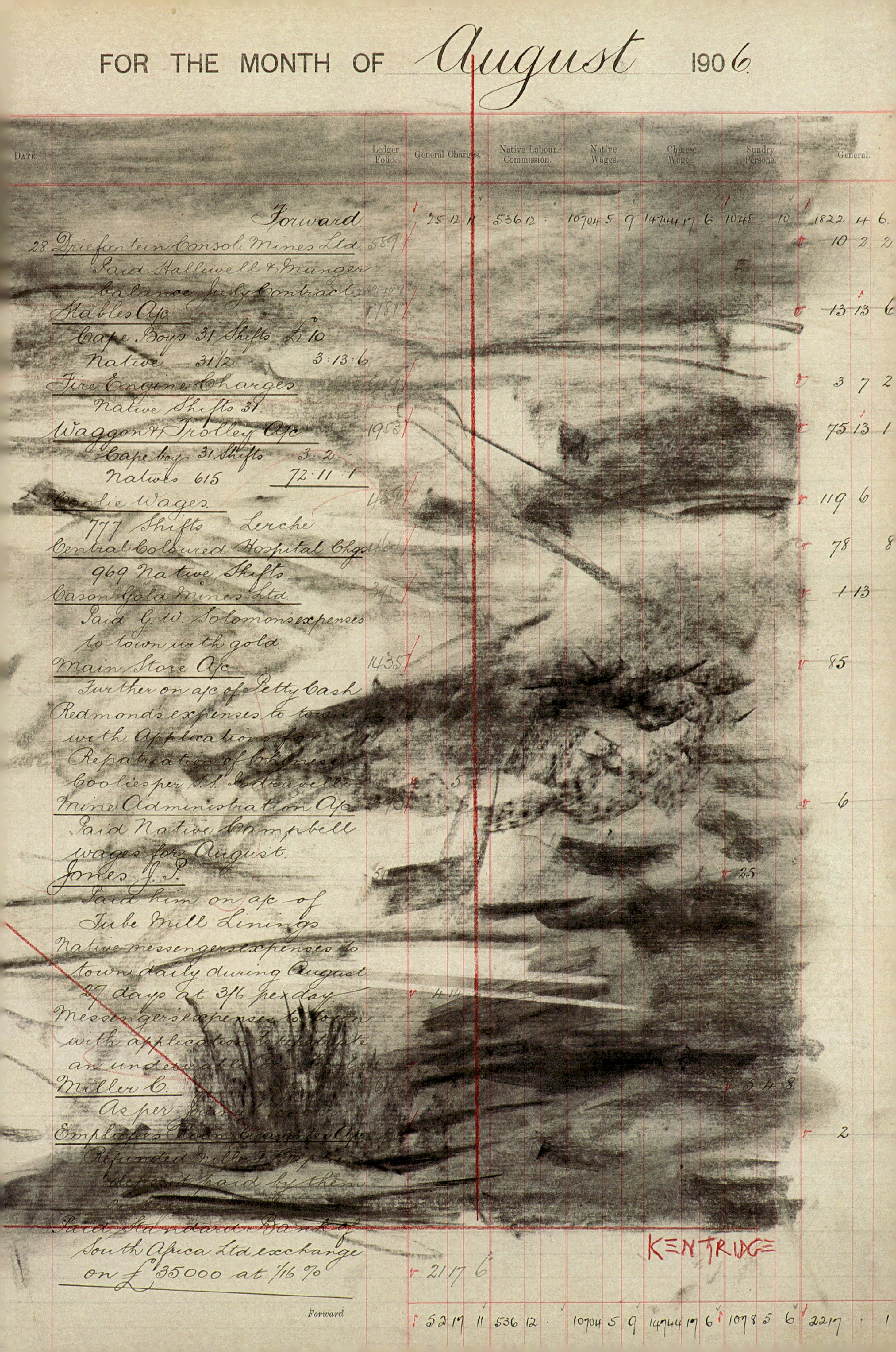

FOR THE MONTH OF August 1906
Date.
Ledger Folio.
General Charges.
Native Labour Commission.
Native Wages.
Chinese Wages.
Sundry Persons.
General.
Forward
28 Driefontein Consol. Mines Ltd.
Paid Halliwell & Munger
balance July Contract
Stables A/c
Cape Boys 31 Shifts
Fire Engine Charges
Native Shifts 31
Waggon & Trolley A/c
Coolie Wages
777 Shifts
Central Coloured Hospital Chgs.
969 Native Shifts
Main Store A/c
Mine Administration A/c
Paid Native Campbell
wages for August
Jones J. S.
Paid him on a/c of
Tube Mill Linings
Native messengers expenses to
town daily during August
Miller C.
Paid Standard Bank of
South Africa Ltd exchange
on £35000
Forward
KENTRIDGE

CENTRAL ADMINISTRATION MINE CASH

Date.		Ledger Folio.	Sundry Persons.	East Rand Proprietary Mines.	Native Advances.	Chinese Advances.	Employees' Accident Assurance Claims Account.	General.
	Forward		866 10 -	86 12 1	373 5 3	992 19 4	358 5 8	29382 12 1
25	Jones B. Owen Ltd. Electrical power supplied £: 4:3 Domestic Water 7:6	456	11 9					
	W. N. L. Assocⁿ R 51/73 £1:14/ B.J. 77/63 11/	992	2 5					
	Edwards T. W. Coal A/c Jany	246	9					
	Received from Hercules Co Ltd as per letter dated 27 April							
	Natives Return Home A/c. Recd from Hercules Co Ltd as per letter dated 27 April	1537						
	The Rand Mutual Assce Coy. Ltd on behalf of Dowdle M. J. £14: 8: 4 Stewart J. 156: 10: - Hodgson J. 28: 3: 10							
	Fisher Bros Main Store A/c	307						
	Koppel Ltd. Sanitary fees March	504						
	Forward		874 1 6	86 12 1	376 5 3	992 19 4	560 7 10	29383 9 7

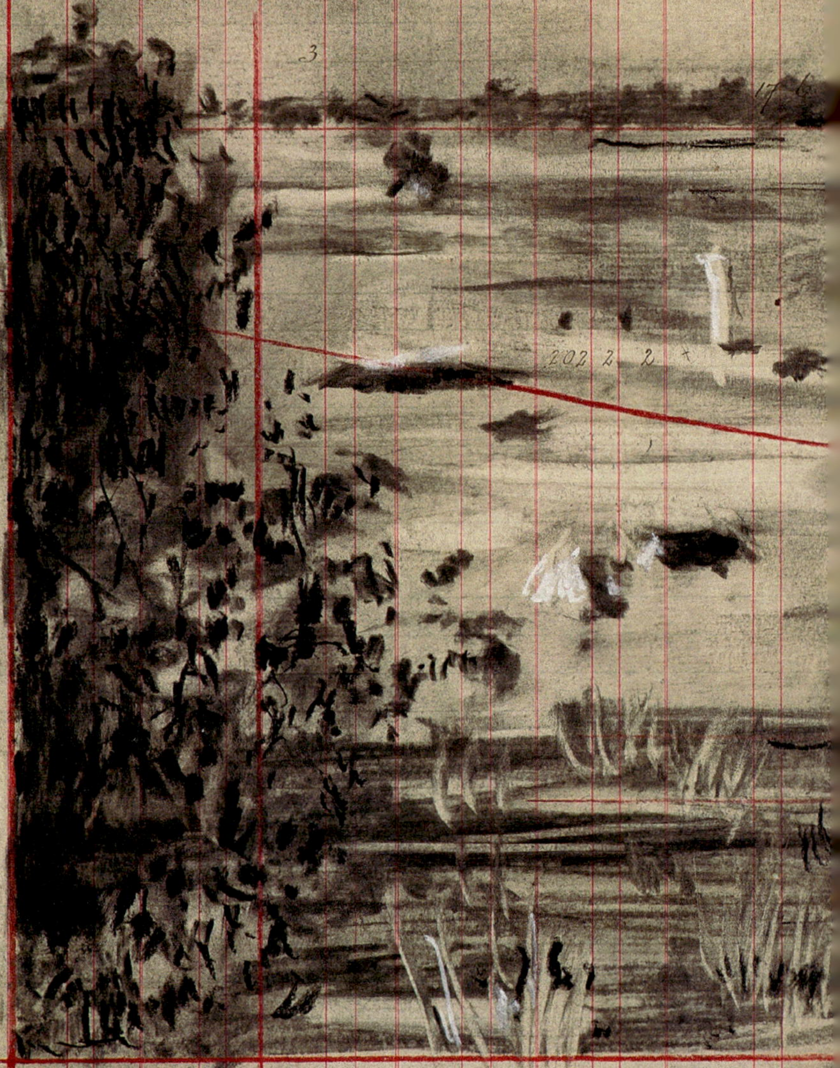

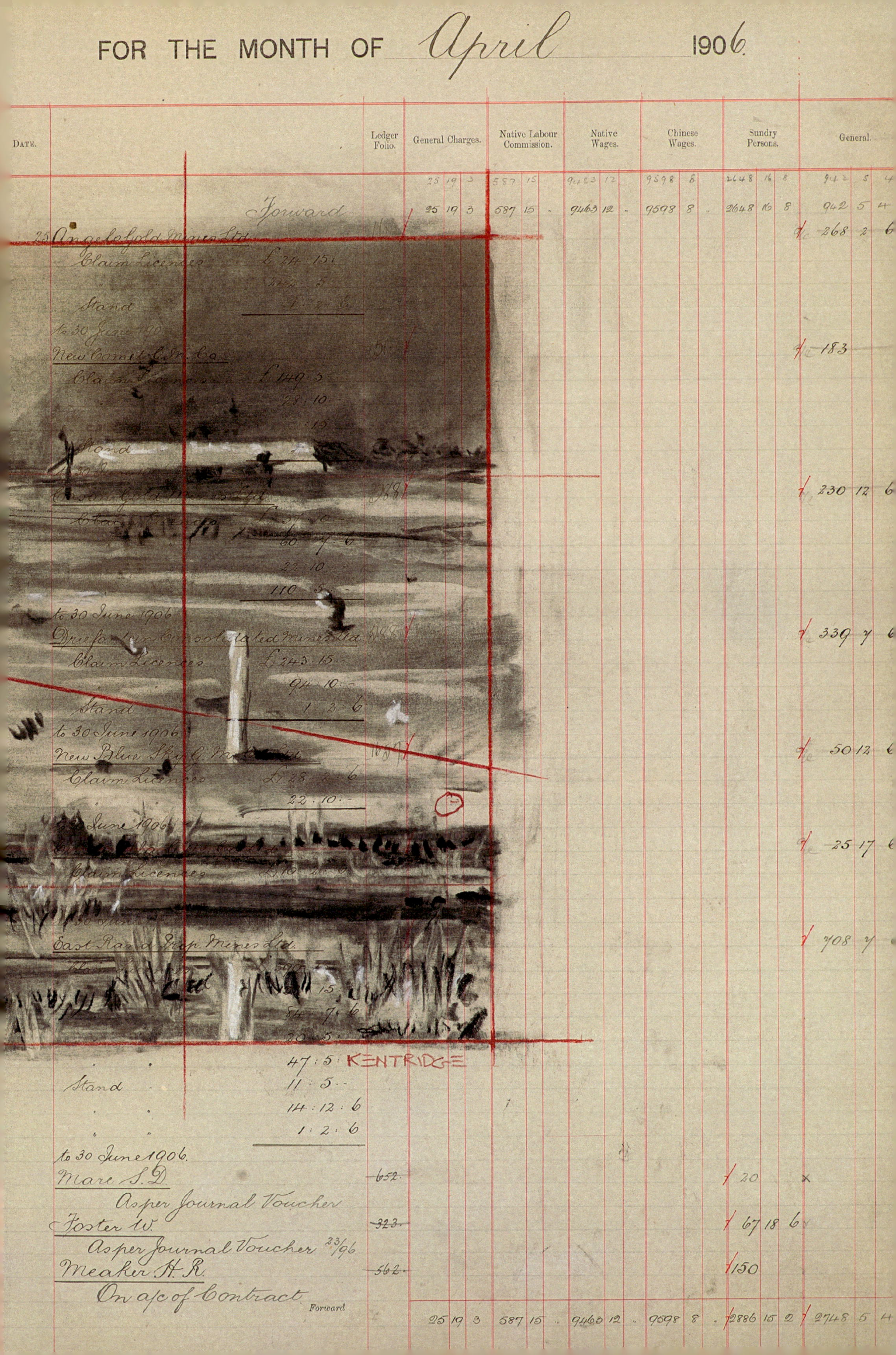

FOR THE MONTH OF April 1906
Date.
Ledger Folio.
General Charges.
Native Labour Commission.
Native Wages.
Chinese Wages.
Sundry Persons.
General.
Forward 25 19 3 587 15 9463 12 9598 8 2648 16 8 942 5 4
25 Angelo Gold Mines Ltd 268 2 6
Claim Licences £24 15
Stand
to 30 June 1906
New Comet G.M. Co. 183
Claim Licences £149 5
230 12 6
to 30 June 1906
Driefontein Consolidated Mines Ltd 339 7 6
Claim Licences £243 15
Stand 1 2 6
to 30 June 1906
New Blue Sky G.M. Co. 50 12 6
Claim Licences
25 17 6
Claim Licences
East Rand Prop. Mines Ltd. 708 7
KENTRIDGE
47 5
Stand 11 5
14 12 6
1 2 6
to 30 June 1906.
Marc S.D. 652 20
As per Journal Voucher
Foster W. 323 67 18 6
As per Journal Voucher 23/96
Meaker H.R. 562 150
On a/c of Contract
Forward 25 19 3 587 15 9463 12 9598 8 2886 15 2 2748 5 4

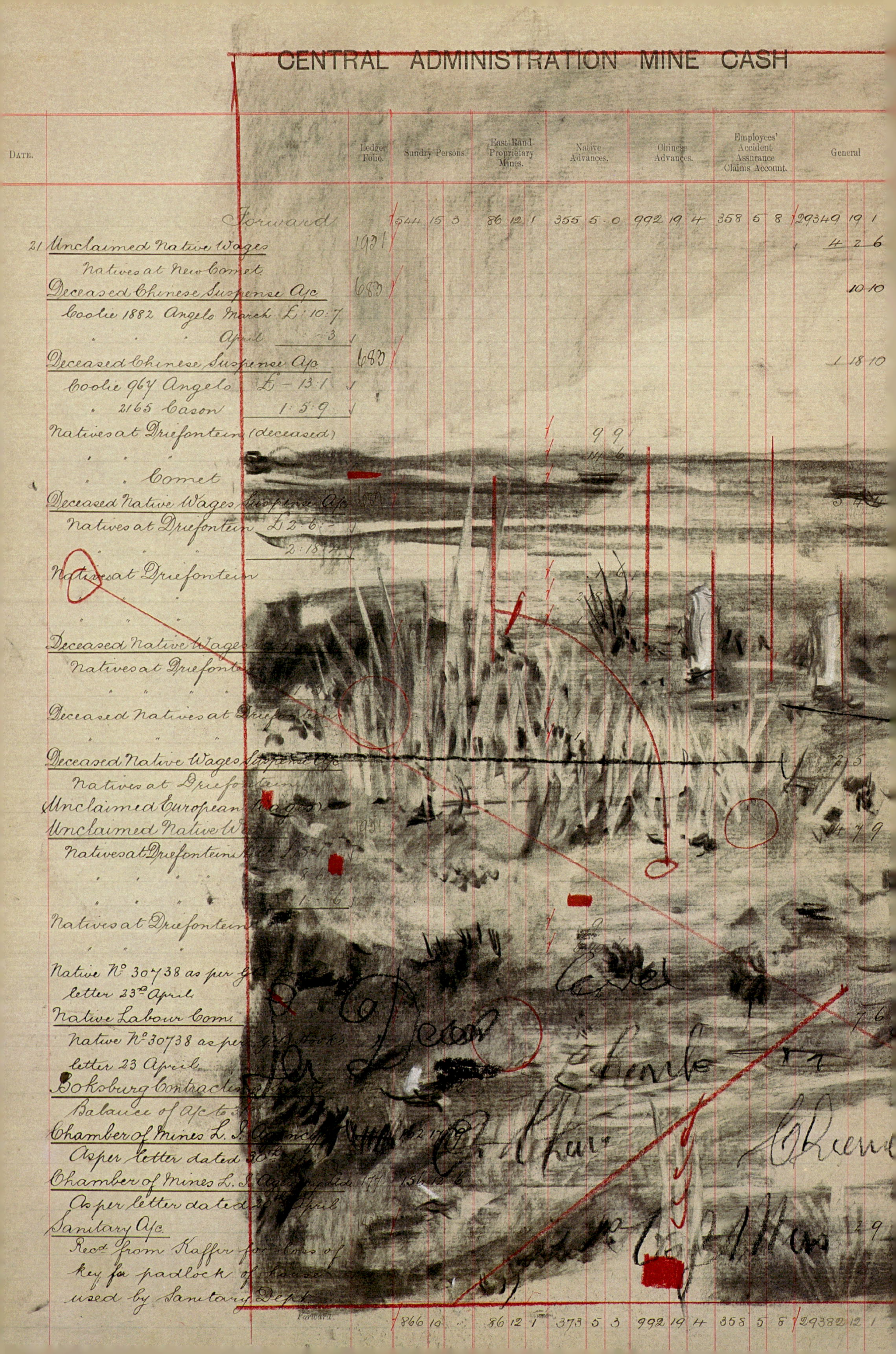
CENTRAL ADMINISTRATION MINE CASH
Date.
Ledger Folio.
Sundry Persons.
East Rand Proprietary Mines.
Native Advances.
Chinese Advances.
Employees' Accident Assurance Claims Account.
General
Forward
Unclaimed Native Wages
Natives at New Comet
Deceased Chinese Suspense A/c
Coolie 1882 Angelo March
April
Deceased Chinese Suspense A/c
Coolie 967 Angelo
2165 Cason
Natives at Driefontein (deceased)
Comet
Deceased Native Wages Suspense A/c
Natives at Driefontein
Natives at Driefontein
Deceased Native Wages
Deceased Natives at Driefontein
Deceased Native Wages Suspense A/c
Unclaimed European Wages
Unclaimed Native Wages
Natives at Driefontein
Natives at Driefontein
Native No 30738 as per
letter 23rd April
Native Labour Com
Native No 30738 as per
letter 23 April
Boksburg Contractors
Balance of A/c to
Chamber of Mines L. I. Agency
As per letter dated 30th
Chamber of Mines L. I. Agency
As per letter dated
Sanitary A/c
Recd from Kaffir for
key for padlock
used by Sanitary Dept
Forward

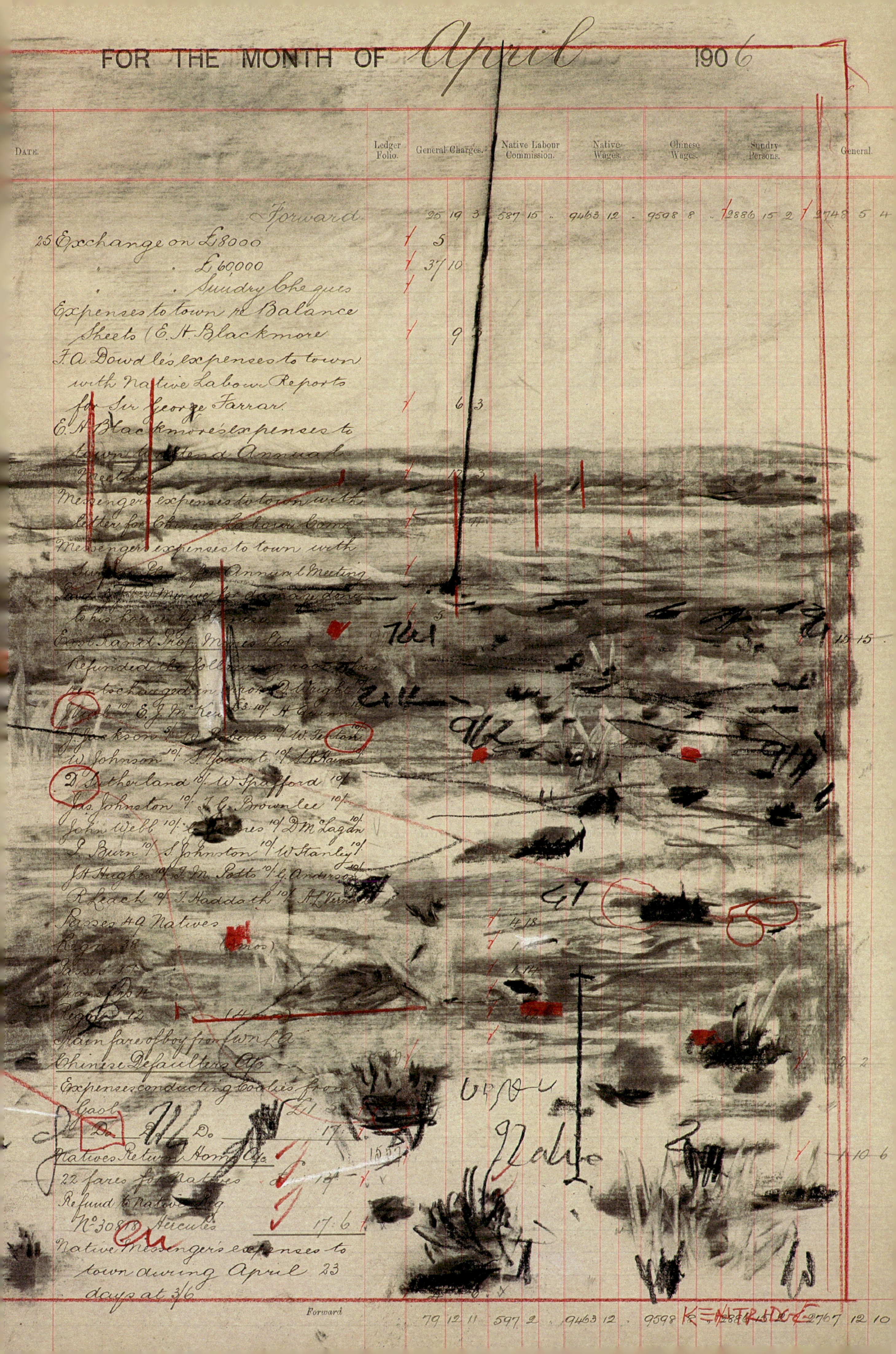

FOR THE MONTH OF April 1906
Date
Ledger Folio.
General Charges.
Native Labour Commission.
Native Wages.
Chinese Wages.
Sundry Persons.
General
Forward
25 Exchange on £8000
£60000
Sundry Cheques
Expenses to town re Balance Sheets (E.H. Blackmore
F.A. Dowdles expenses to town with Native Labour Reports for Sir George Farrar
E.H. Blackmores expenses to
Messengers expenses to town with
Chinese Defaulters A/c
Expenses conducting Coolies from Gaol
Natives Return Home A/c
22 fares for Natives
Native Messengers expenses to town during April 23 days at 3/6
Forward
KENTRIDGE

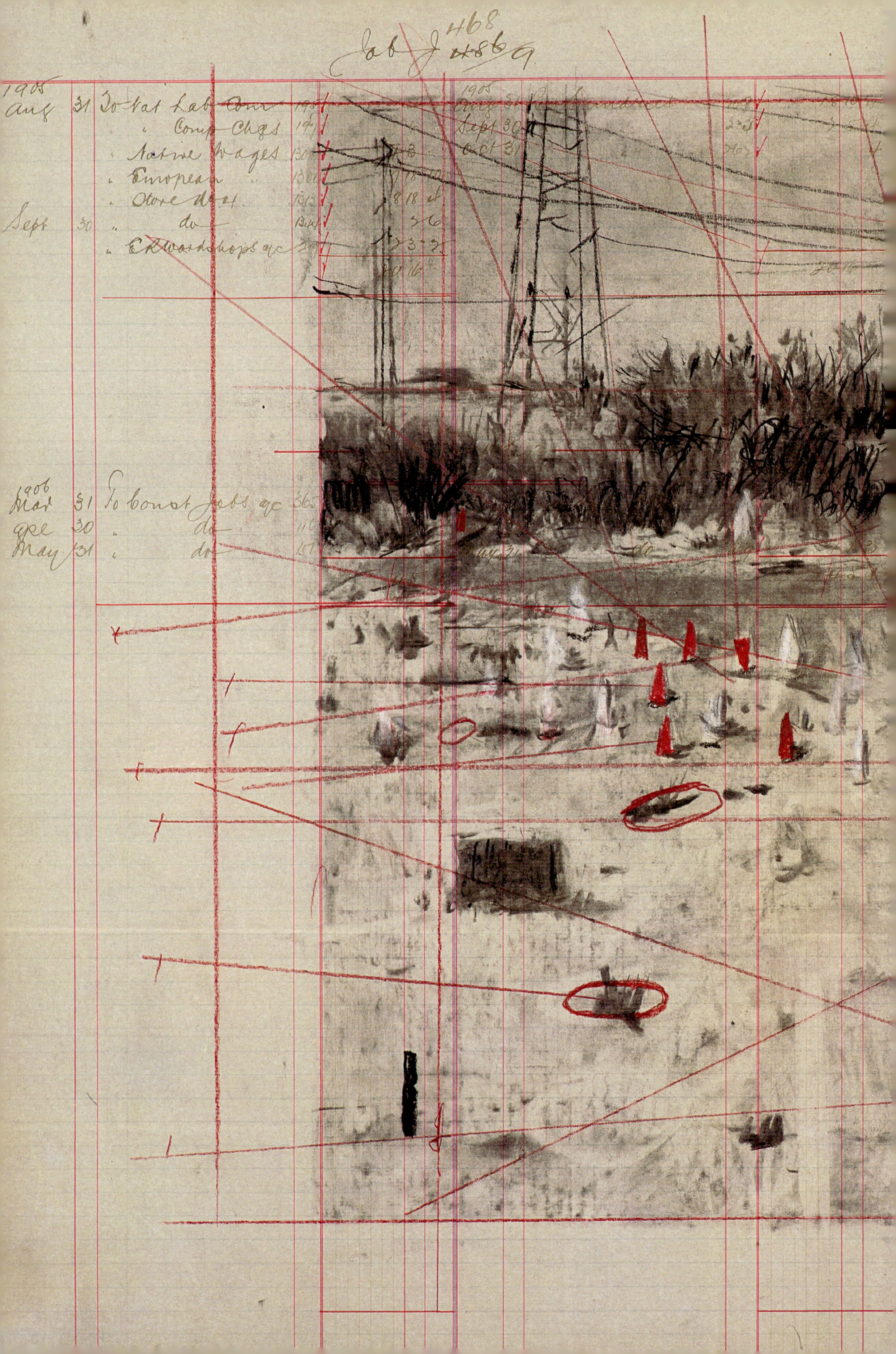
Job 468
1905
Aug 31
Comp Chgs
Native Wages
European
Sept 30
1906
Mar 31 To Const Jobs &c
May 31

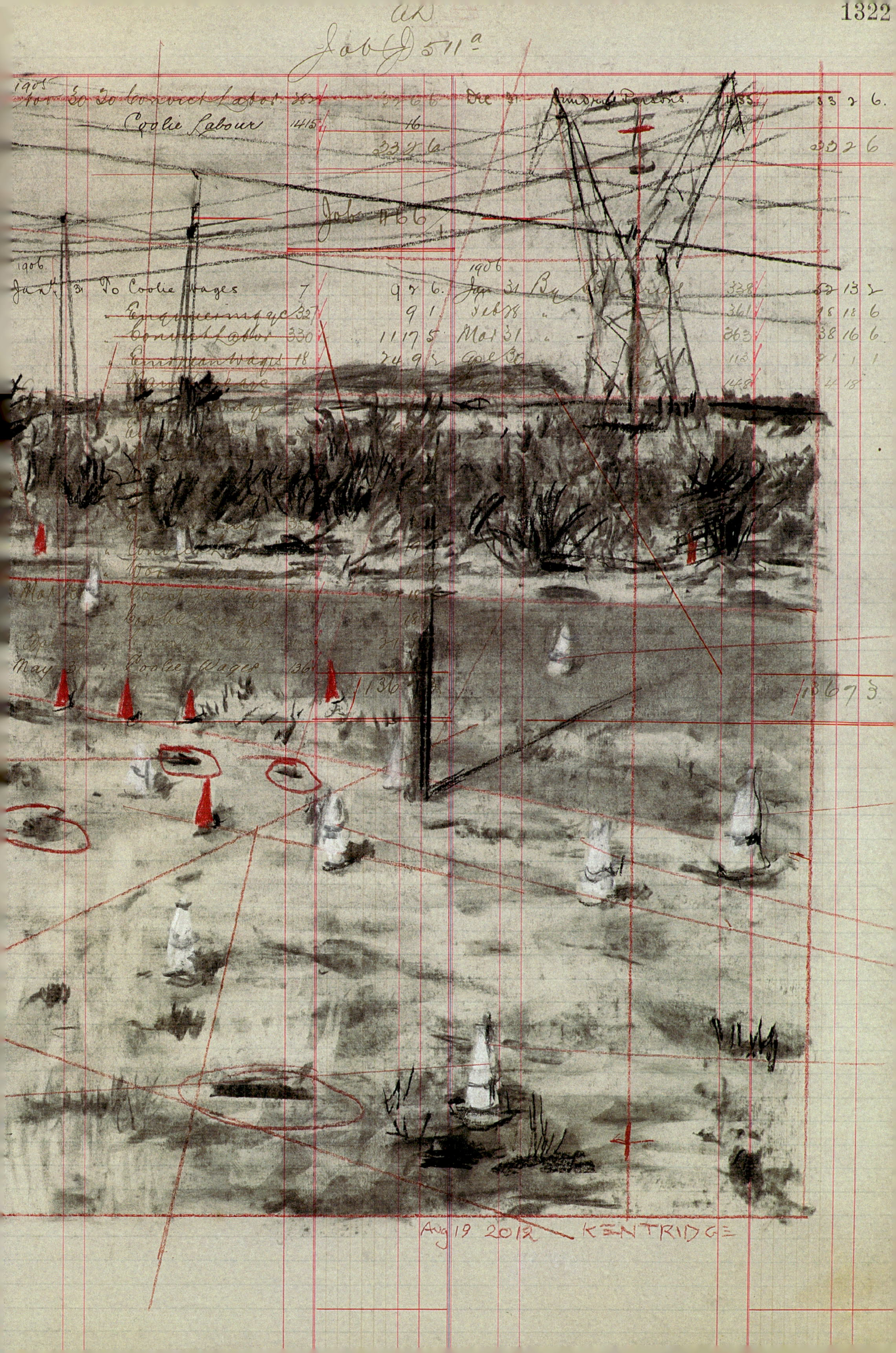
1322
Job 511a
Convict Labor
Coolie Labour
To Coolie wages
Coolie Wages
Aug 19 2012 KENTRIDGE

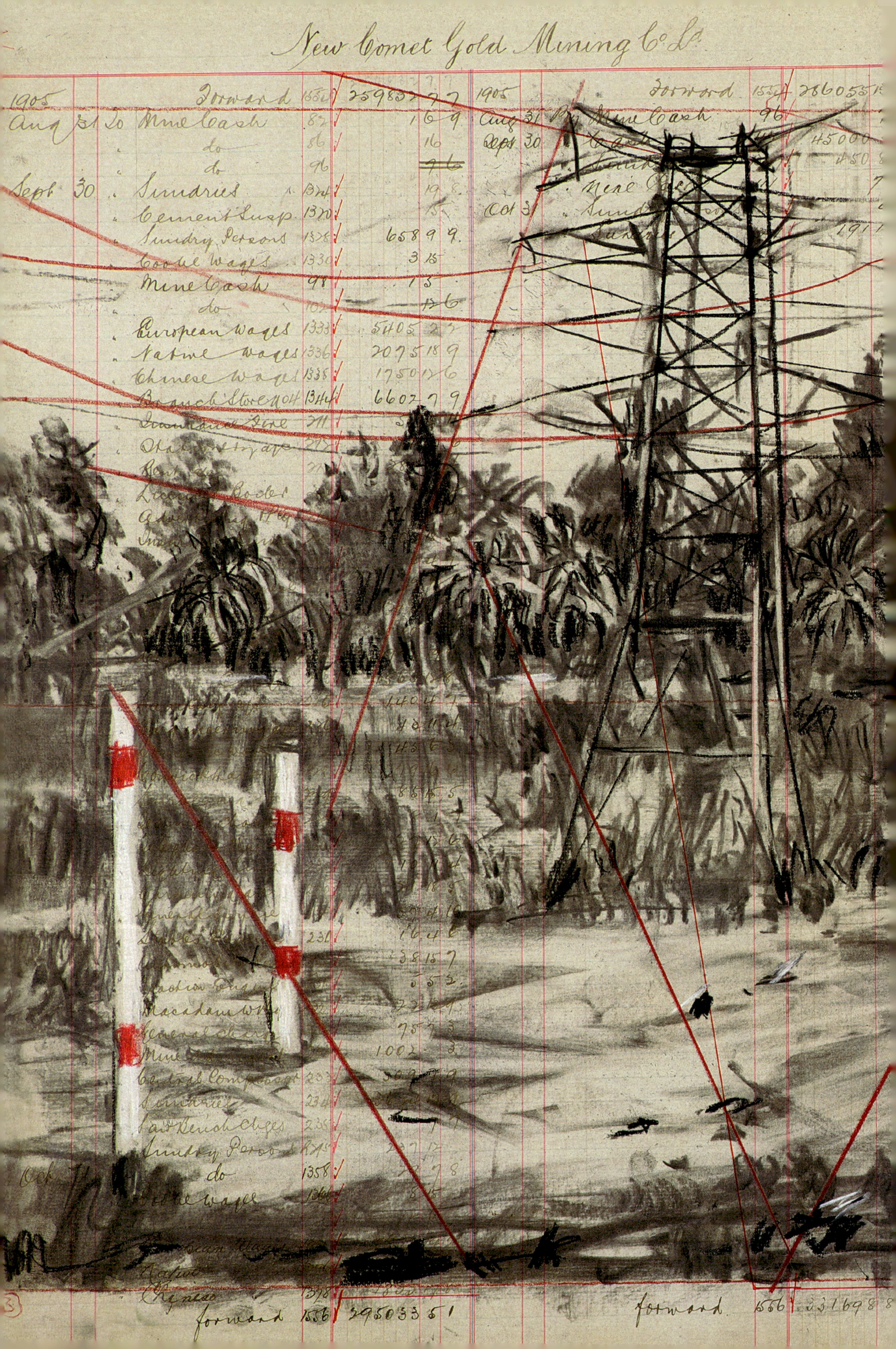
New Comet Gold Mining Co Ld
1905 Forward
Aug 31 To Mine Cash
Sept 30 Sundries
Cement Susp
Sundry Persons
Coolie Wages
Mine Cash
European Wages
Native Wages
Chinese Wages
forward

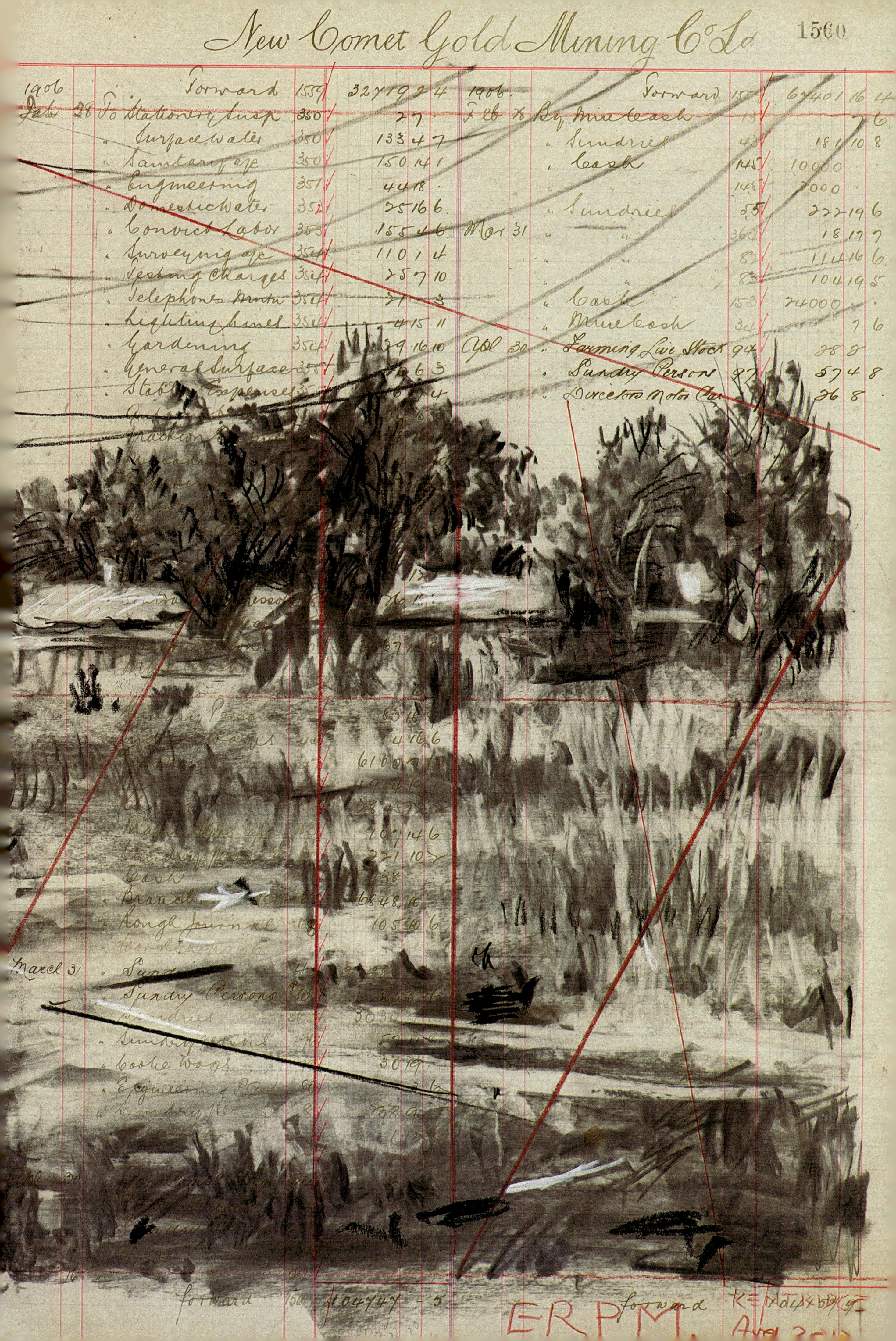
New Comet Gold Mining Co Ld
1560
ERPM

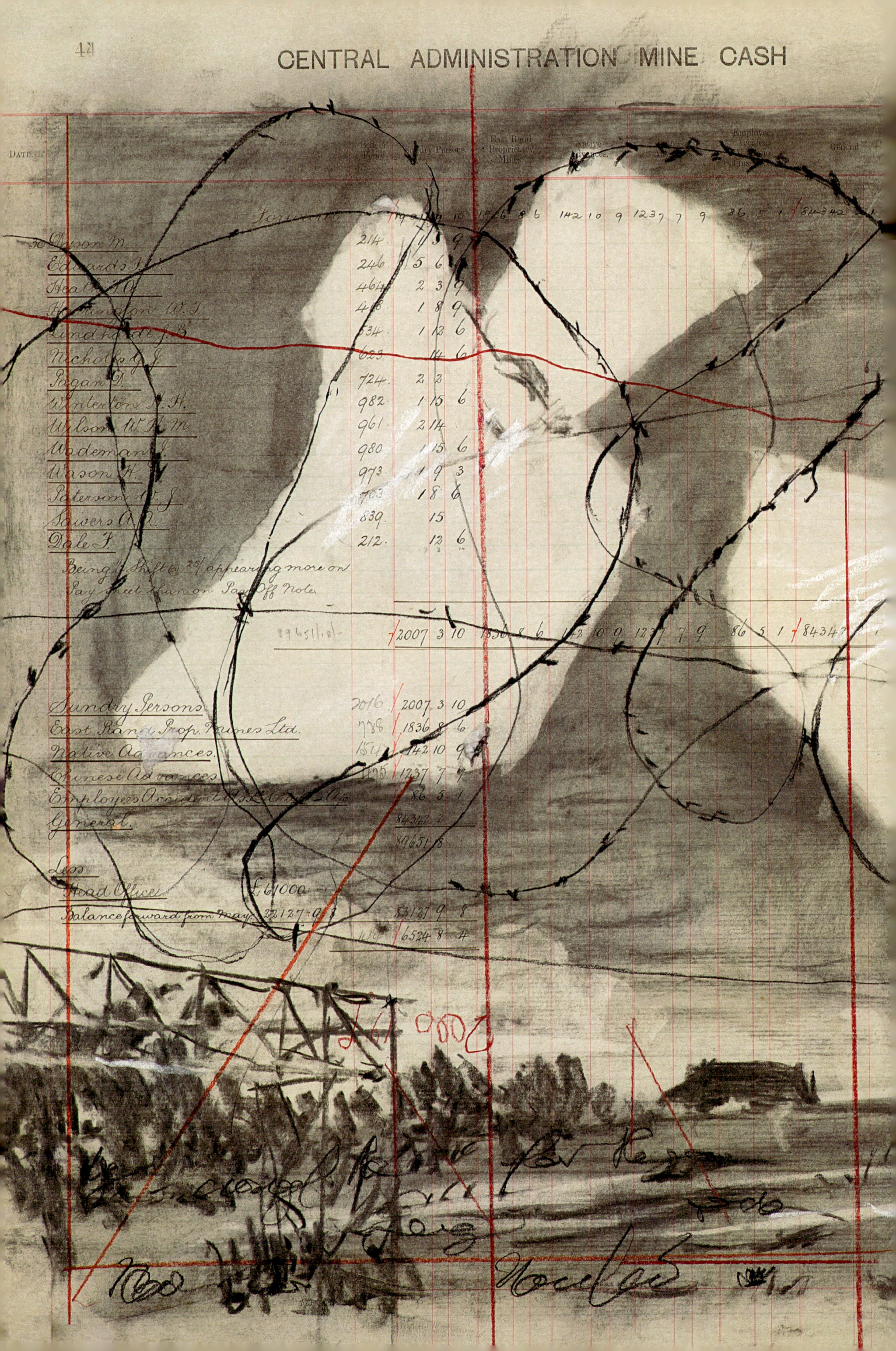
CENTRAL ADMINISTRATION MINE CASH

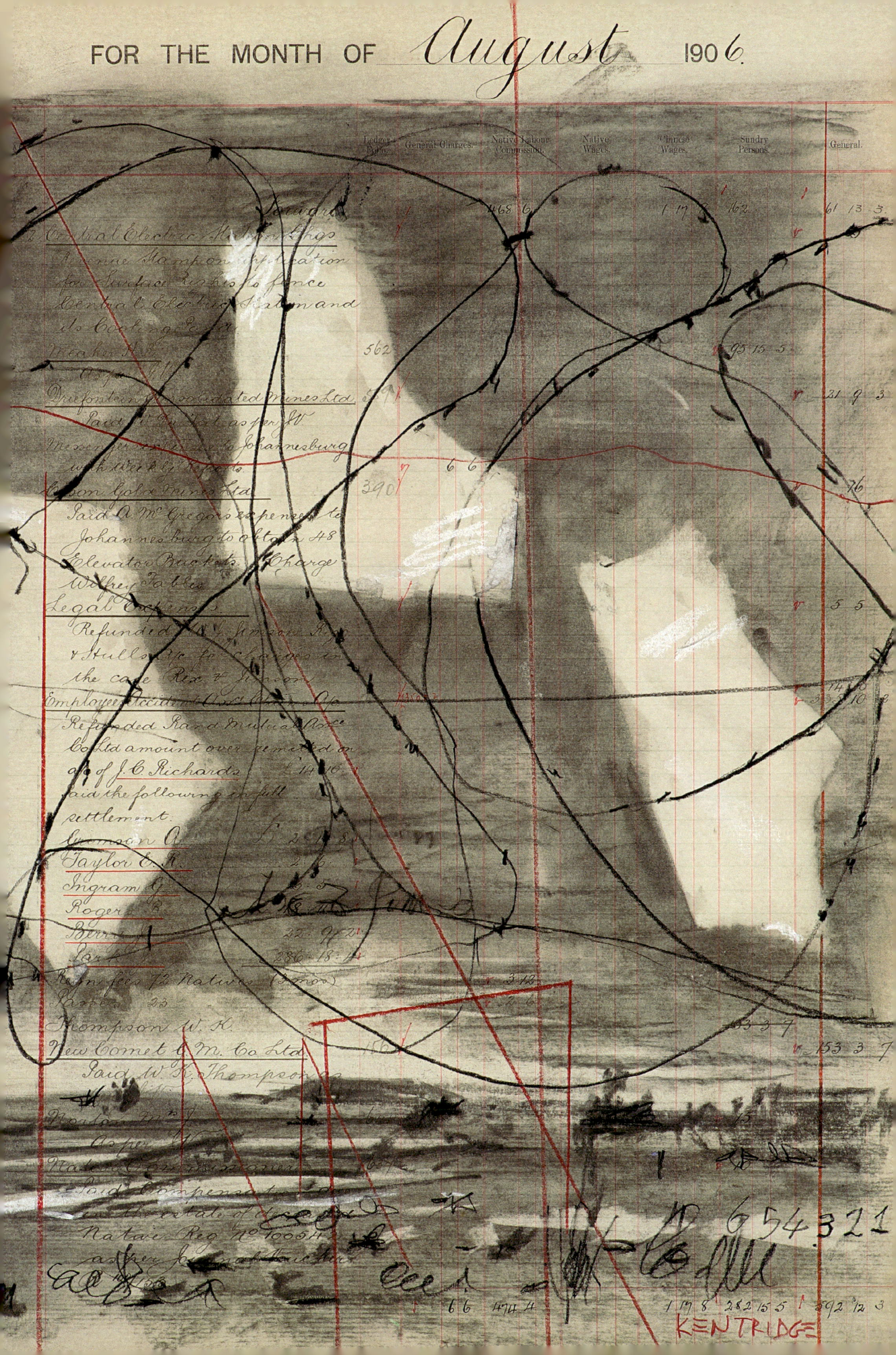
FOR THE MONTH OF August 1906
KENTRIDGE

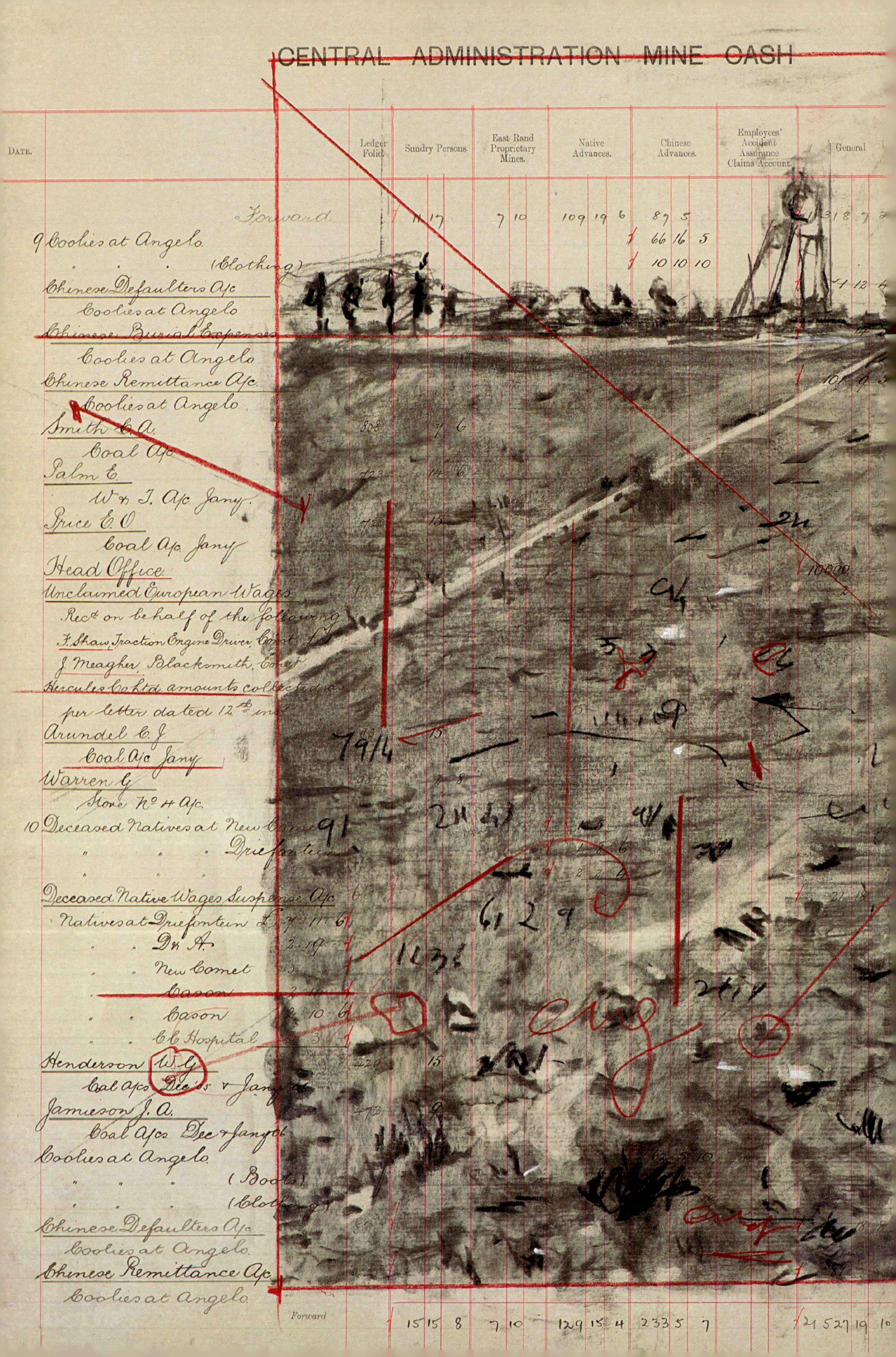
CENTRAL ADMINISTRATION MINE CASH
Date.
Ledger Folio.
Sundry Persons.
East Rand Proprietary Mines.
Native Advances.
Chinese Advances.
Employees' Accident Assurance Claims Account.
General
Forward
9 Coolies at Angelo
(Clothing)
Chinese Defaulters A/c
Coolies at Angelo
Chinese Burial Expenses
Coolies at Angelo
Chinese Remittance A/c
Coolies at Angelo
Smith C.A.
Coal A/c
Salm E.
W & T. A/c Jany
Price E.O.
Coal A/c Jany
Head Office
Unclaimed European Wages
Recd on behalf of the following
F. Shaw, Traction Engine Driver
J. Meagher, Blacksmith
Hercules Co Ltd amounts collected
per letter dated 12th inst
Arundel C.J.
Coal A/c Jany
Warren G.
Store No 4 A/c
10 Deceased Natives at New Comet
Driefontein
Deceased Native Wages Suspense A/c
Natives at Driefontein
D & A
New Comet
Cason
Cason
CC Hospital
Henderson W.G.
Coal A/cs Dec & Jany
Jamieson J.A.
Coal A/cs Dec & Jany 06
Coolies at Angelo
(Boots)
(Clothing)
Chinese Defaulters A/c
Coolies at Angelo
Chinese Remittance A/c
Coolies at Angelo
Forward

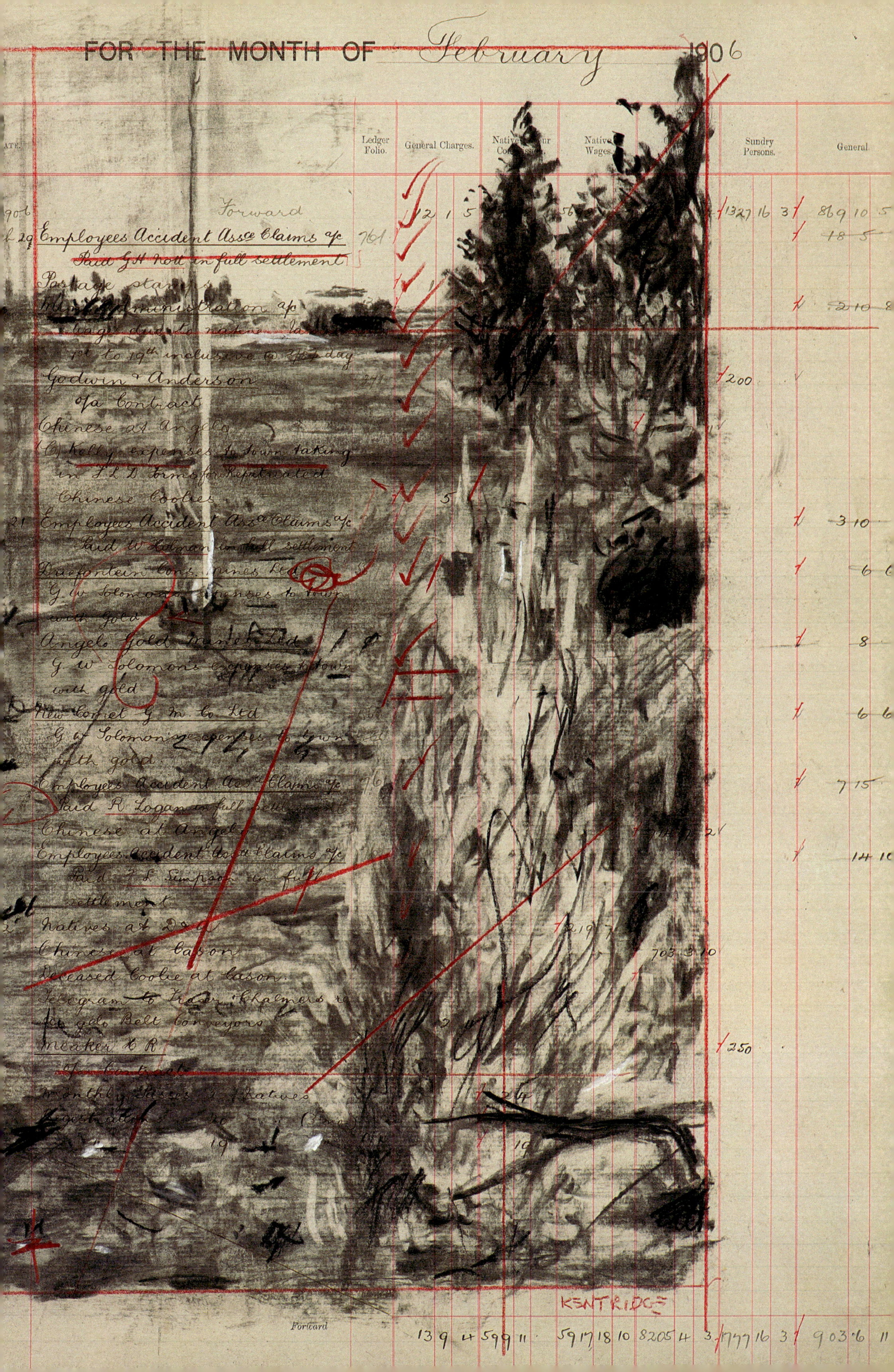

FOR THE MONTH OF February 1906
Ledger Folio.
General Charges.
Sundry Persons.
General
Forward
Employees Accident Assn Claims a/c
Godwin & Anderson
a/c Contract
Chinese Coolies
Chinese at Angelo
Deceased Coolie at Gason
New Comet G. M. Co. Ltd.
Forward
KENTRIDGE

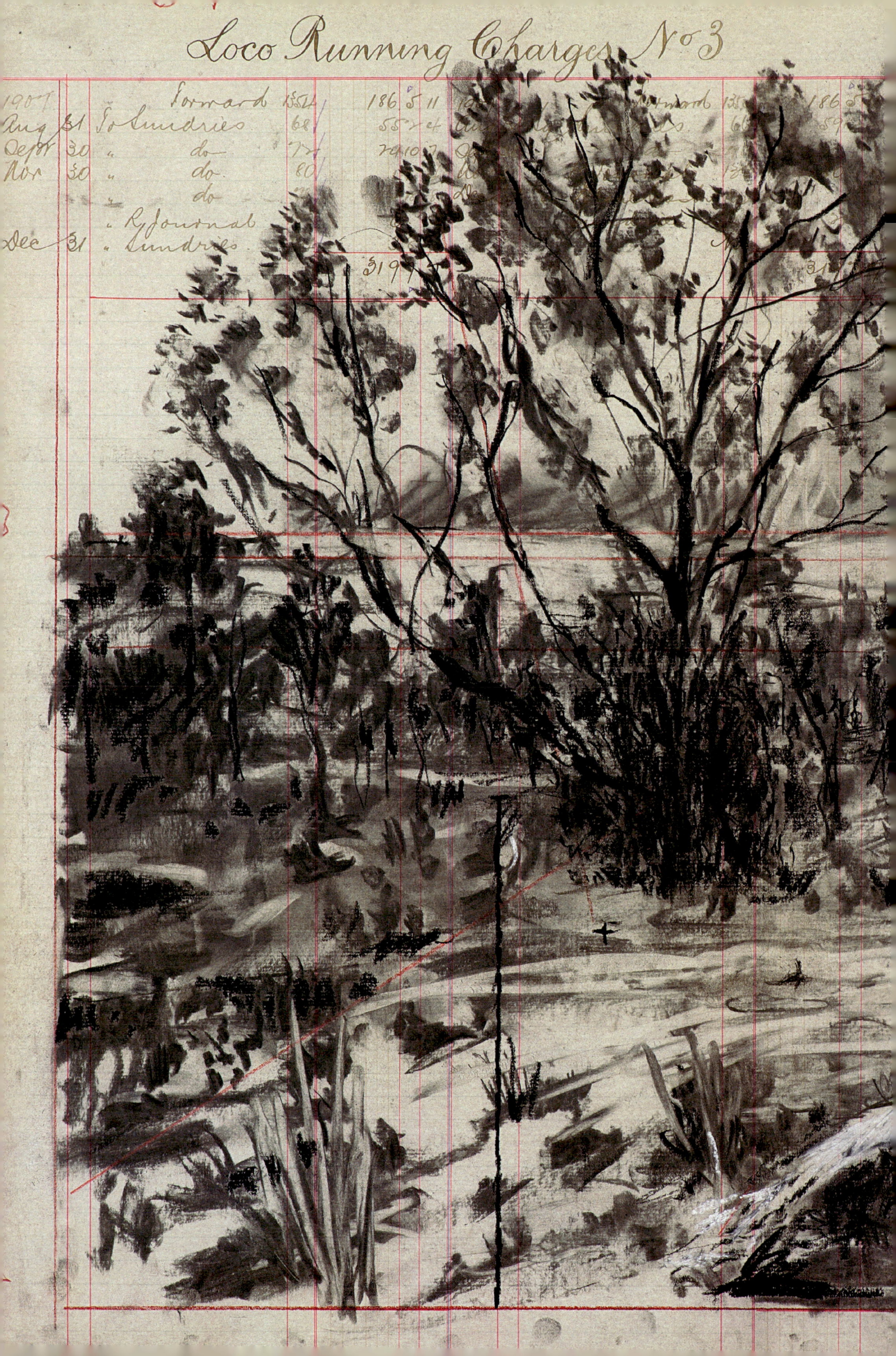
Loco Running Charges No 3
1907 Forward
Aug 31 To Sundries
Sept 30 " do
Nov 30 " do
" do
" Ry Journal
Dec 31 " Sundries

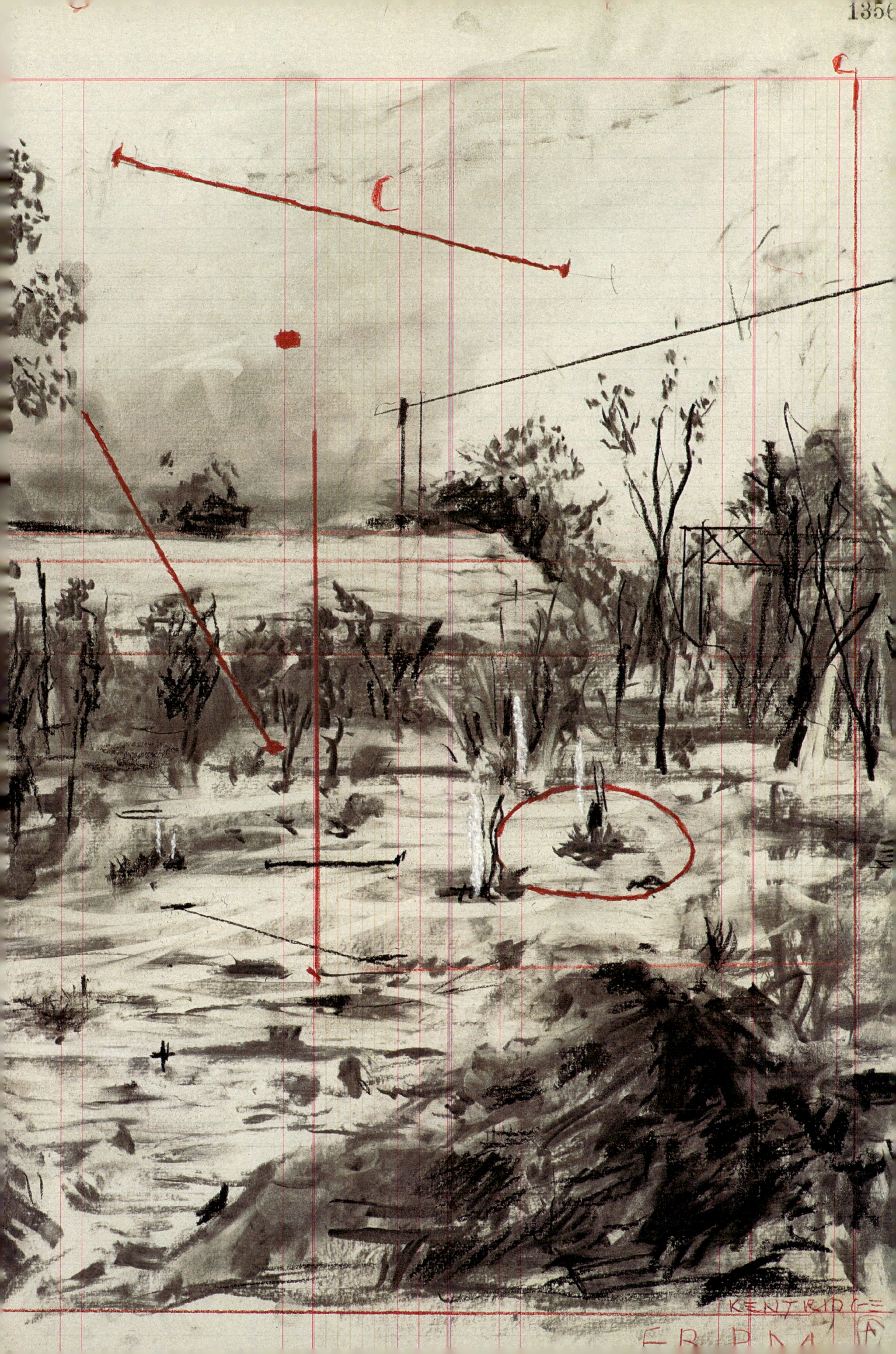
KENTRIDGE

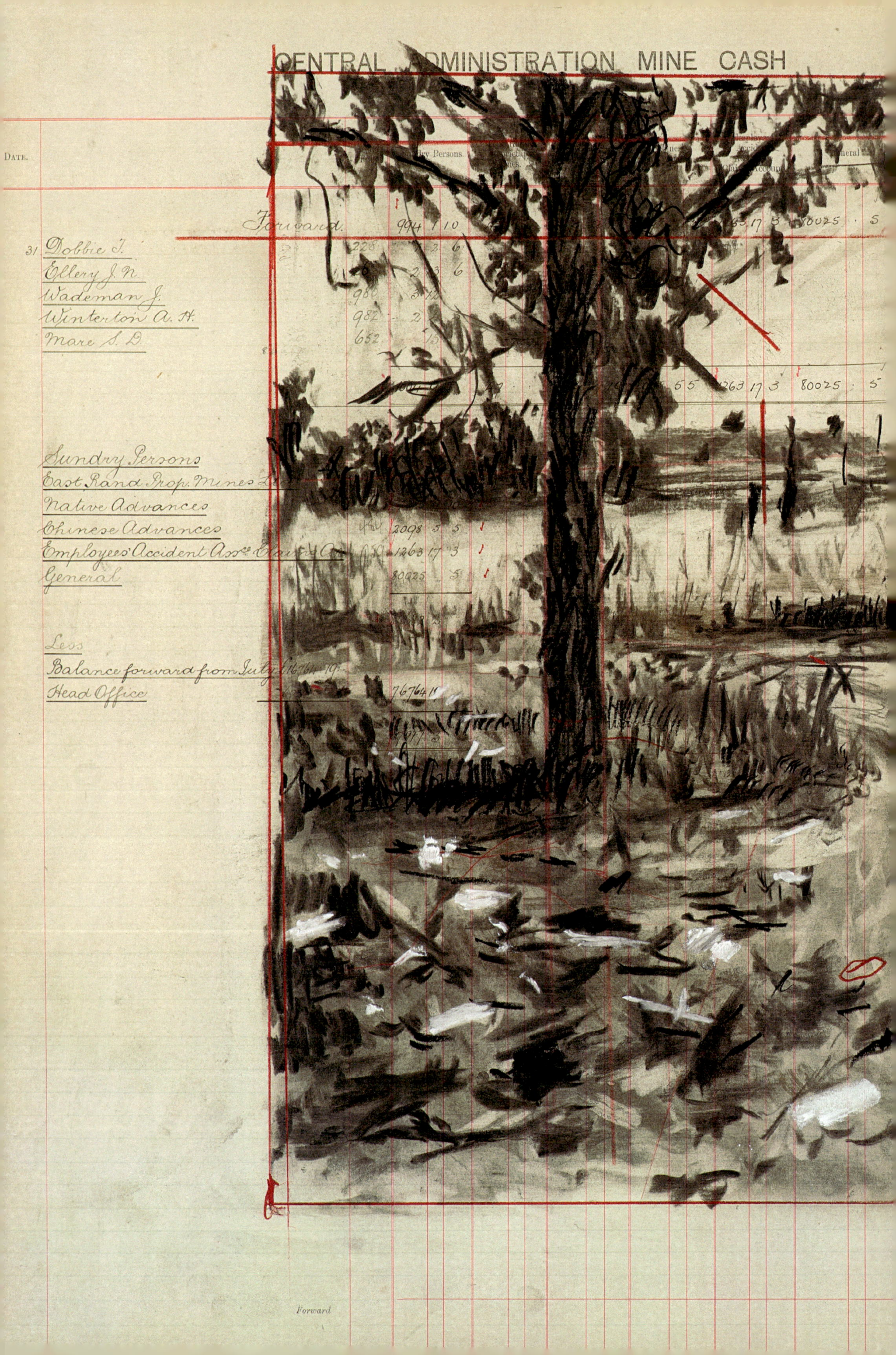
CENTRAL ADMINISTRATION MINE CASH
DATE.
Forward
Dobbie J.
Ellery J. N.
Wademan J.
Winterton A. H.
Mare S. D.
Sundry Persons
East Rand Prop. Mines Ld
Native Advances
Chinese Advances
Employees' Accident Assn.
General
Less
Balance forward from July
Head Office
Forward

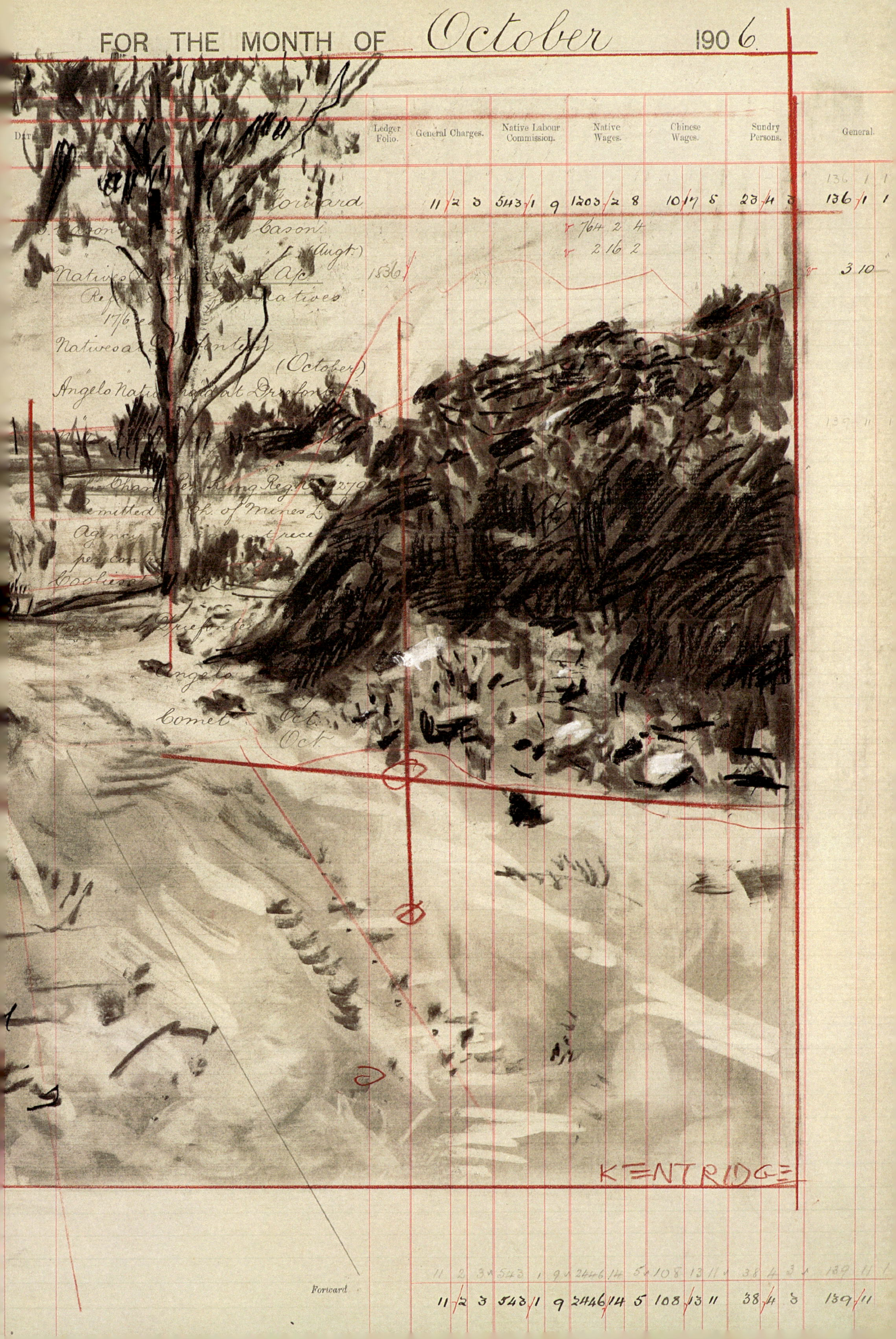
FOR THE MONTH OF October 1906
Ledger Folio.
General Charges.
Native Labour Commission.
Native Wages.
Chinese Wages.
Sundry Persons.
General.
Forward
KENTRIDGE

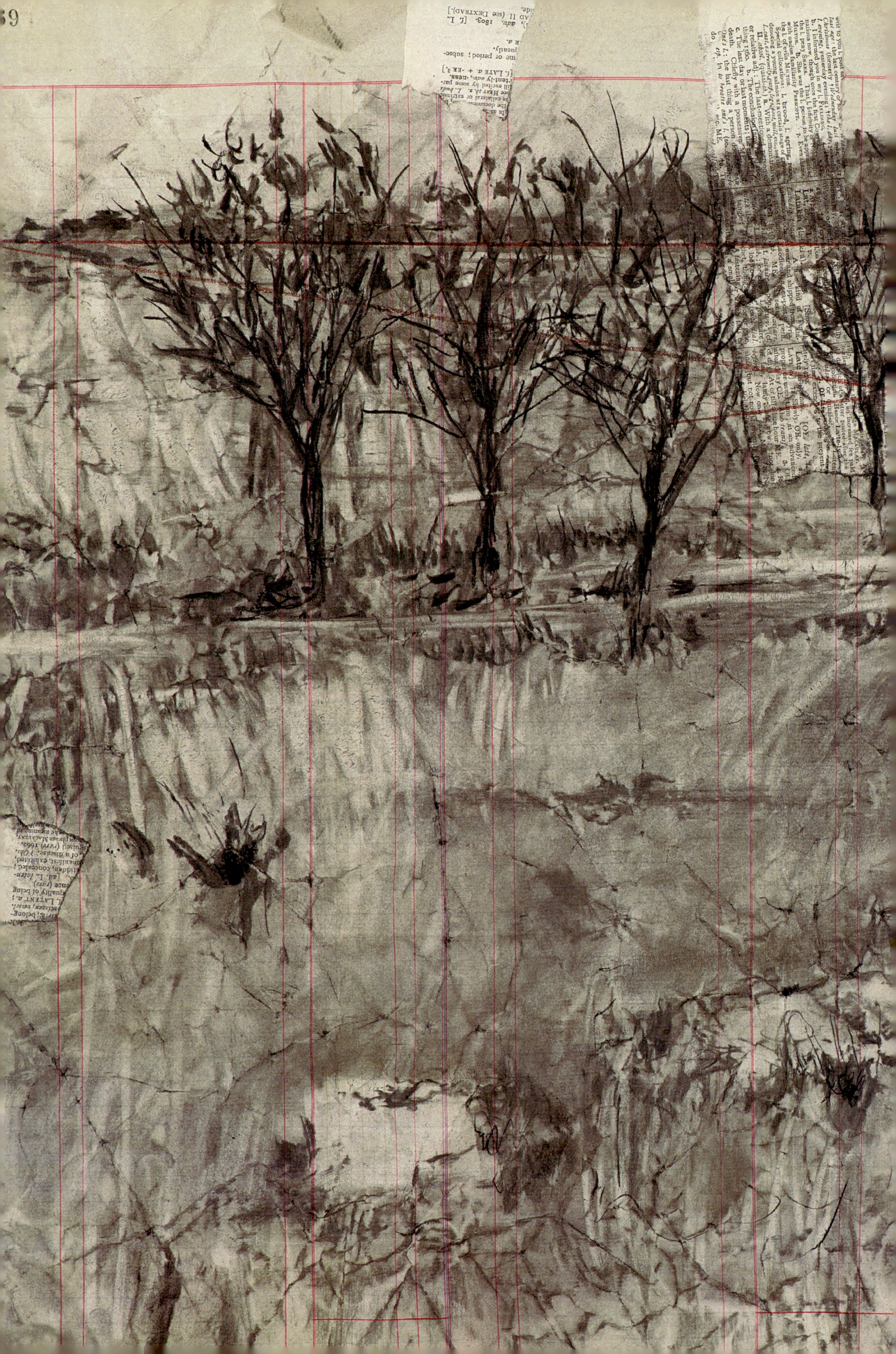

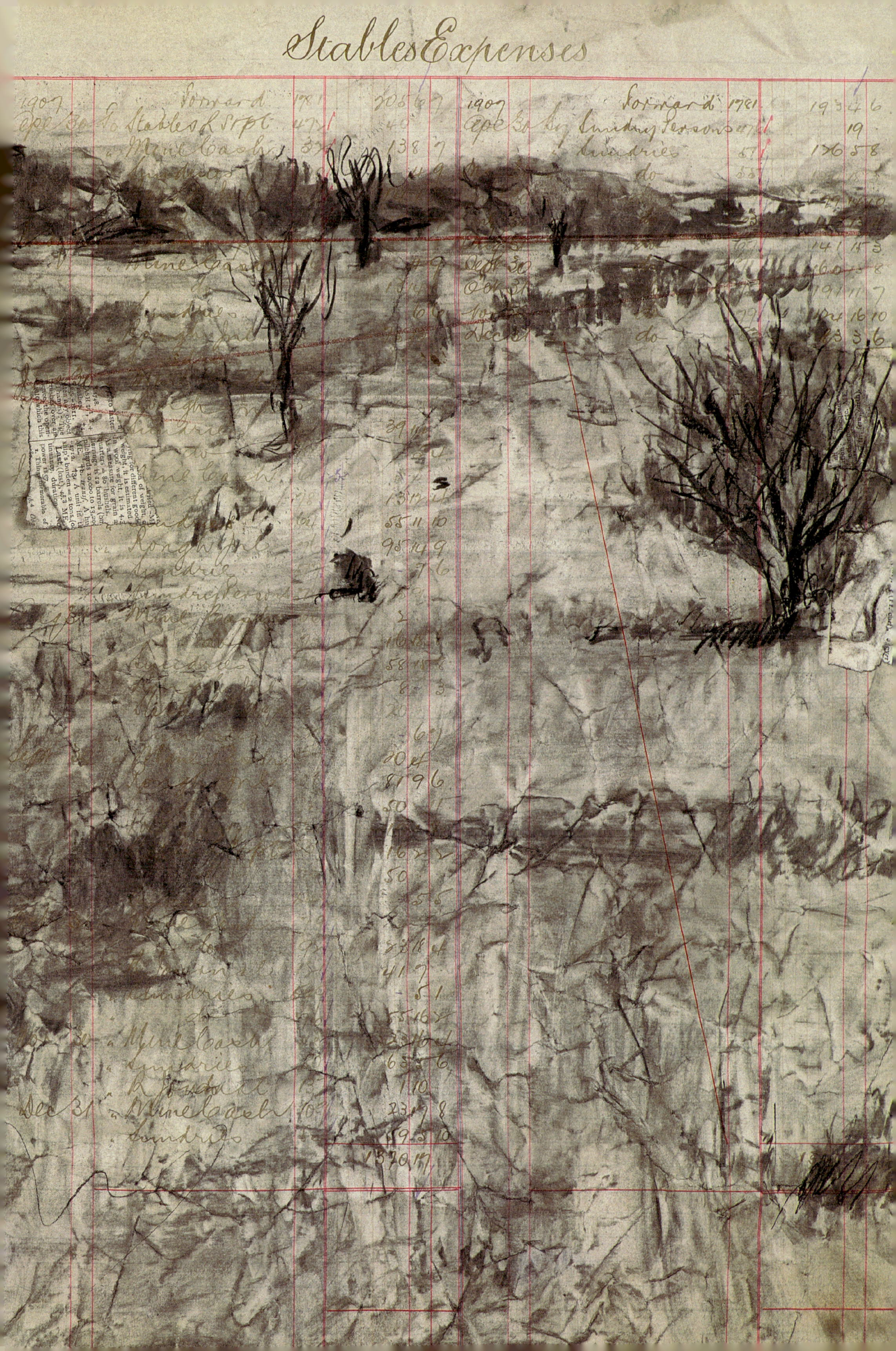
Stables Expenses

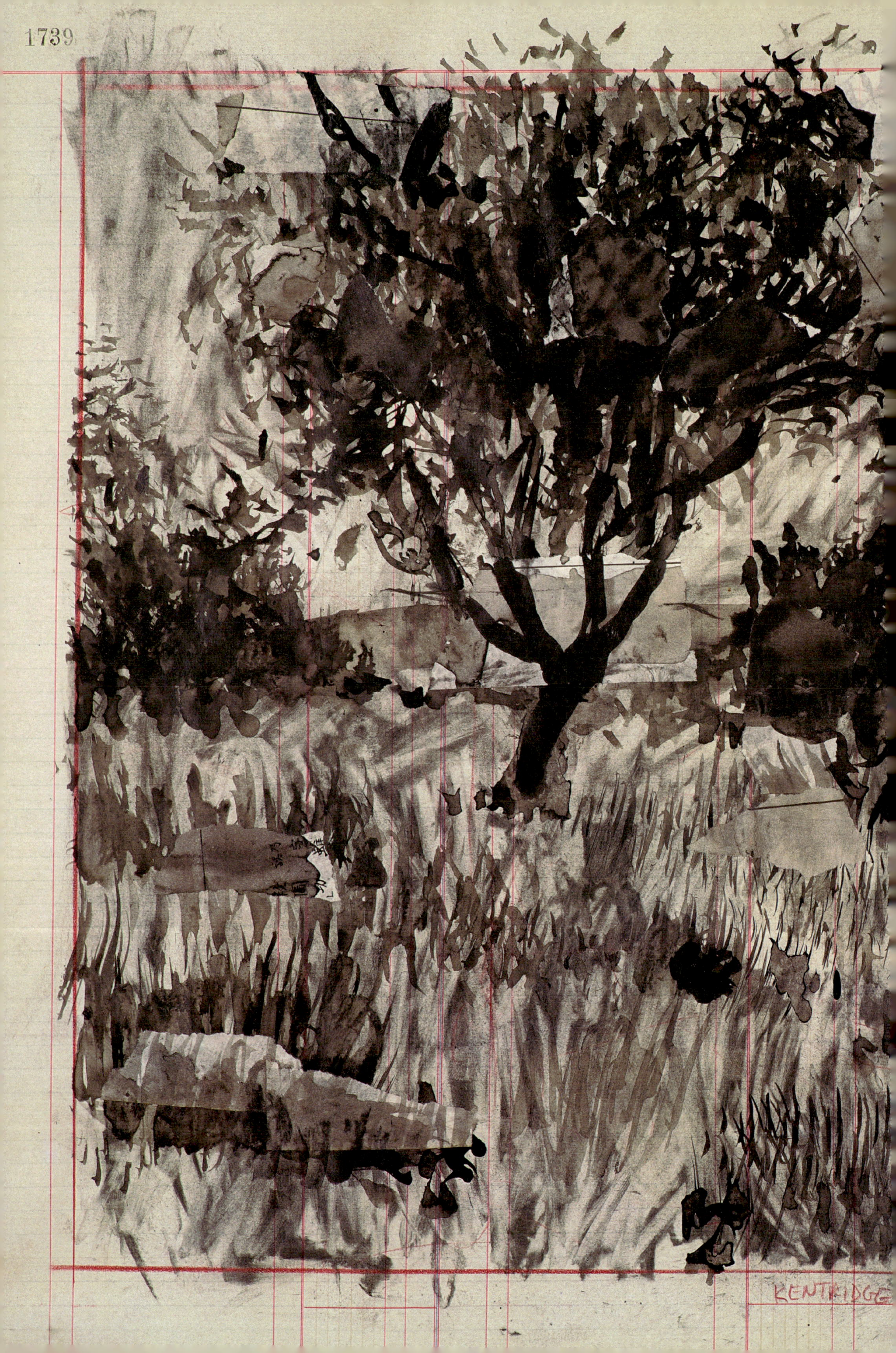
1739
KENTRIDGE

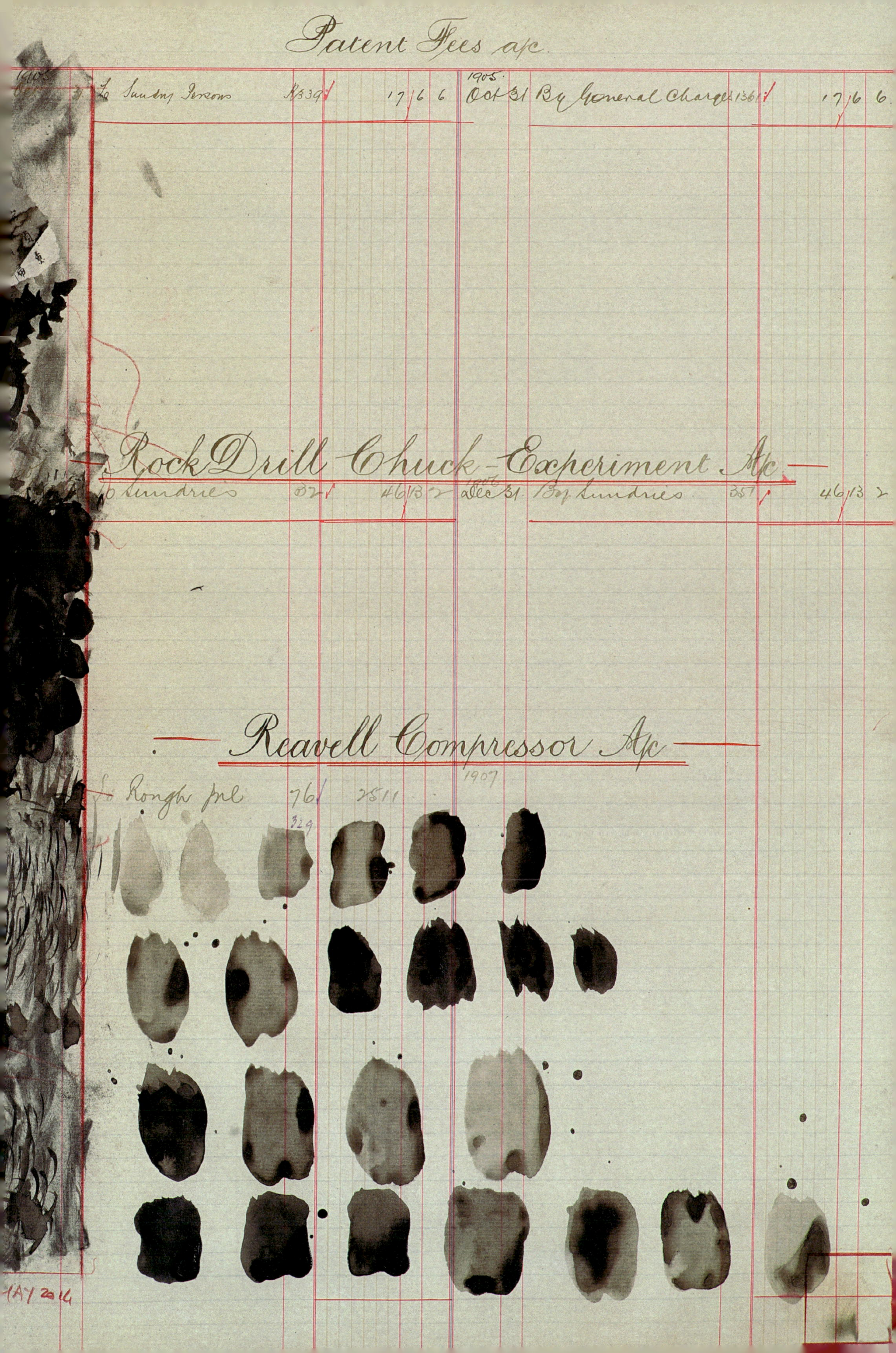
Patent Fees a/c
1905 To Sundry Persons 339 17 6 6
1905 Oct 31 By General Charges 136 17 6 6
Rock Drill Chuck – Experiment A/c
To Sundries 32 46 13 2
1906 Dec 31 By Sundries 351 46 13 2
Reavell Compressor A/c
To Rough Jnl 76 25 11
1907
329

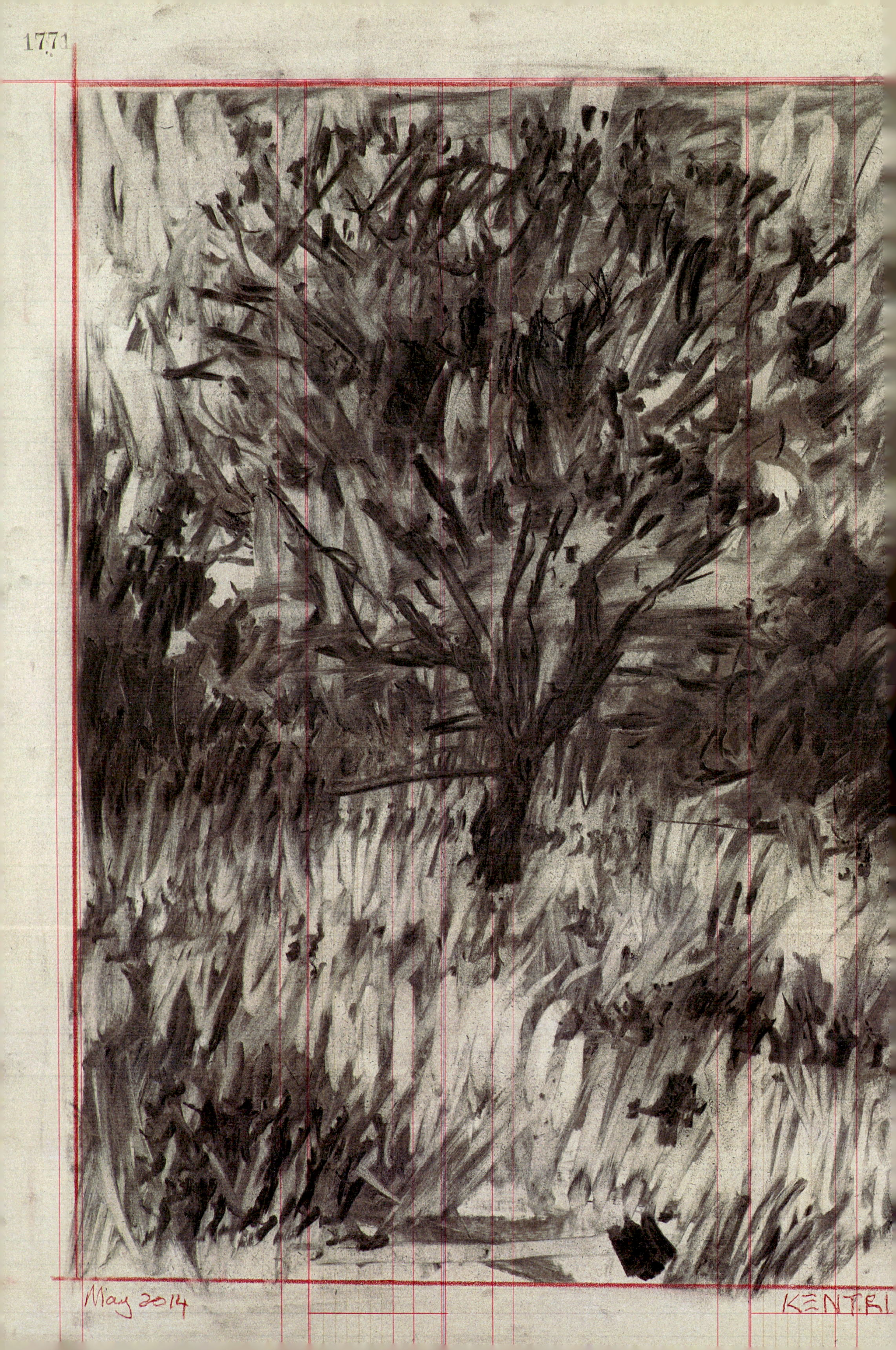
1771
May 2014
KENTRI

Stables Expenses

Date		Particulars	Fol.	£	s.	d.	Date		Particulars	Fol.	£	s.	d.
1905		Forward	1779	814	18	1	1905		Forward	1779	746	16	3
Sept	30	To Domestic Water	227		17	2	Sept	30	By Sundries	231	68	19	1
Oct	[illegible]	" Sundry Persons	1353		14		Oct	31	" Sundry Persons	1366		4	6
		" European Wages	373	4	.				" Sundries	357	78	7	1
		" Rent a/c	240	19	11	6	Nov	30	" Sundry Persons a/c	1400		5	.
		" East Rand Wkshops	243	2	6	9			" Sundries	769	65	16	7
		" Grass Cutting	246	1	8				" Sundry Persons	1435		7	6
		" Col Labor Comm[n]	248	1		7	Dec[r]	31	" Sundries	309	45	8	8
		" Col Comp[d] Chgs	250	1	12	6							
		" Domestic Water Supply	334		13	.							
		" Mine Cash	171	6									
		" Licences Bbg M	341	1									
		" Branch Store No 4	1386	40	.	3							
Nov	30	" European Wages	1407	3	19	.							
		" Rent a/c	764	19	11	6							
		" Licenses B'burg M'pal a/c	766	1									
		" Grass cutting a/c	765	1	8								
		" Sundries	772	3	8	7							
		" "	1477	71	13	10							
		" Sundry Persons a/c	16	7	2	.							
		" Mine Cash	137	5	13	6							
		" "	138	1	.	.							
		" E R Workshops	376	6	18	1							
		" Domestic Water	788		13	.							
		" Rent a/c	791	19	11	6							
		" Licences	793	1									
		" European Wages	1443	3	19								
		" Grass cutting a/c	795	1	4	6							
		" Sundries	798	3	7	4							
		" E. Rand Workshops	307	7	4	10							
		" Domestic Water Supply	306		5	3							
		" Branch Store No 4	1437	35	11	7							
		" Mine Cash	148		18	3							
		" " "	154	6	5	.							
				1035	19	2					1035	19	2

Date		Particulars	Fol.	£	s.	d.	Date		Particulars	Fol.	£	s.	d.
1906 Jan	1	To Waggon Trolley	316	1	4		1906 Jan[y]	31	By Sundry Persons a/c	5	1	6	6
		" Rent a/c	319	19	11	6			" Sundries	332	75	1	3
		" Mine Cash	15	9	1	.	Feb	28	" Sundry Persons	42		8	6
		" Licences B M	320	4									
		" Grass Cutting	321	1	4	6							
		" Col Lab Chges	323	5	6	7							
		" Domestic Water	326	1	10	3							
		" East R Wkshps	334	6	13	5							
		" European Wages	17	5	15								
		" Branch Store No 4	7	27	1	6							
Feb	28	" Mine Cash	10	14	5								
		" Sundry Persons a/c	40	4	10								
		" European Wages	47	1	19								
		" Branch Store No 4	66	17	19	5							
		" Col Lab. Chges.	344	4	19	5							
		" Stationery a/c	345		14	5							
		forward	1781	120	18				forward	1781	76	16	3

CENTRAL ADMINISTRATION MINE CASH

Date.		Ledger Folio.	Sundry Persons.	East Rand Proprietary Mines.	Native Advances.	Chinese Advances.	Employees' Accident Assurance Claims Account.	General
	Forward							

FOR THE MONTH OF 190

Date.		Ledger Folio.	General Charges.	Native Labour Commission.	Native Wages.	Chinese Wages.	Sundry Persons.	General.
	Forward							

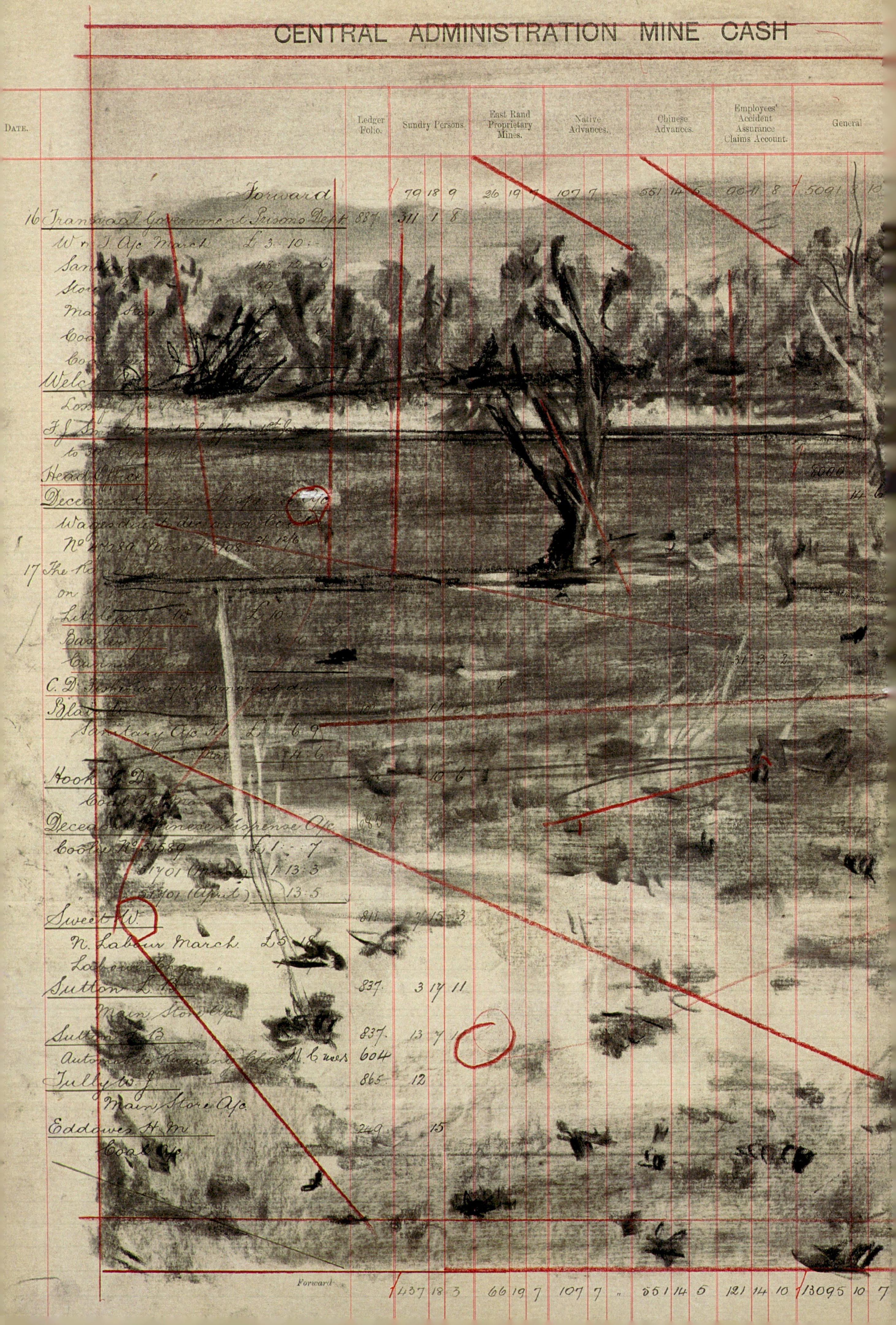
CENTRAL ADMINISTRATION MINE CASH
DATE.
Ledger Folio.
Sundry Persons.
East Rand Proprietary Mines.
Native Advances.
Chinese Advances.
Employees' Accident Assurance Claims Account.
General
Forward
Forward

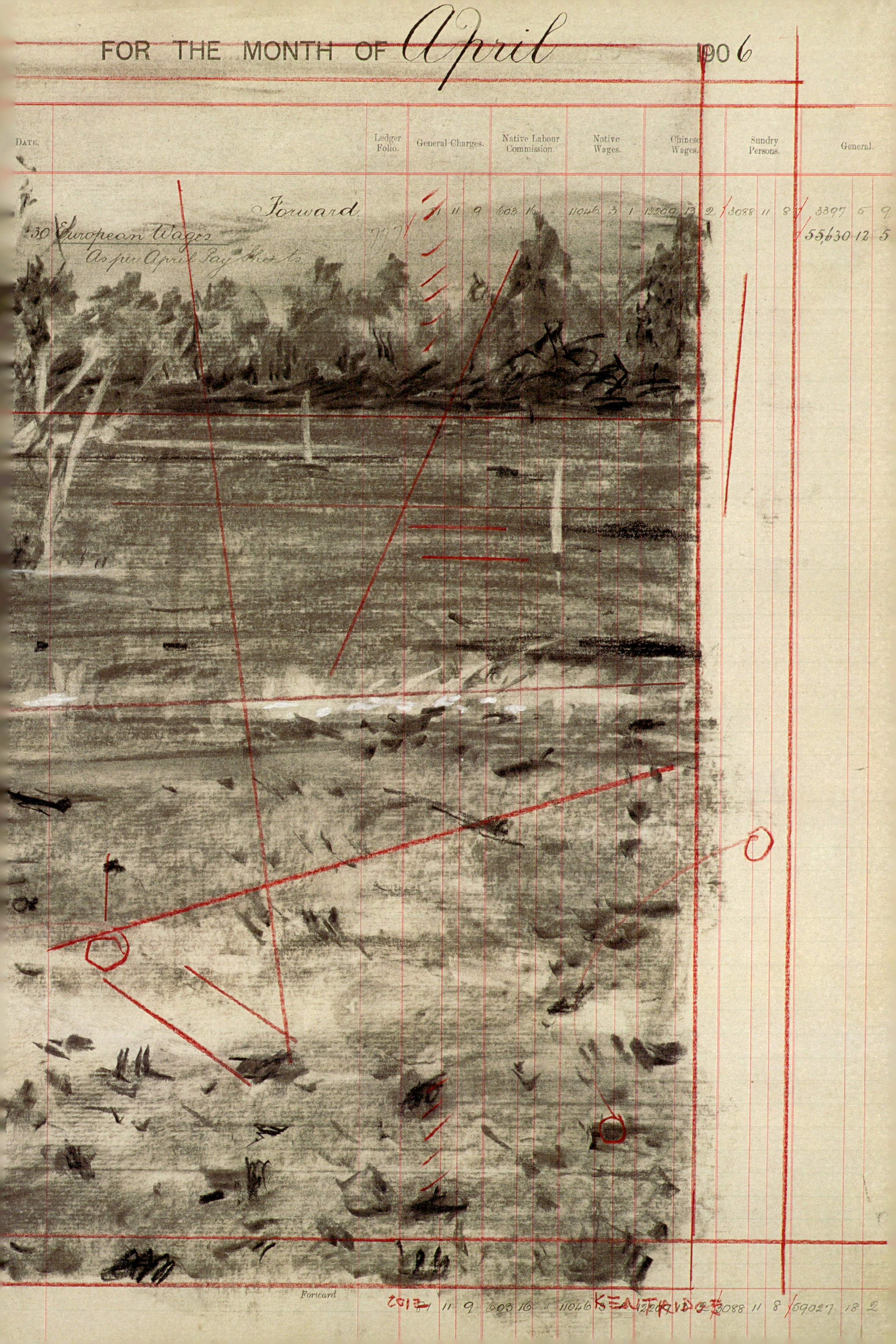
FOR THE MONTH OF April 1906
Date.
Ledger Folio.
General Charges.
Native Labour Commission.
Native Wages.
Chinese Wages.
Sundry Persons.
General.
Forward
30 European Wages
As per April Pay Sheets
Forward
KENTRIDGE

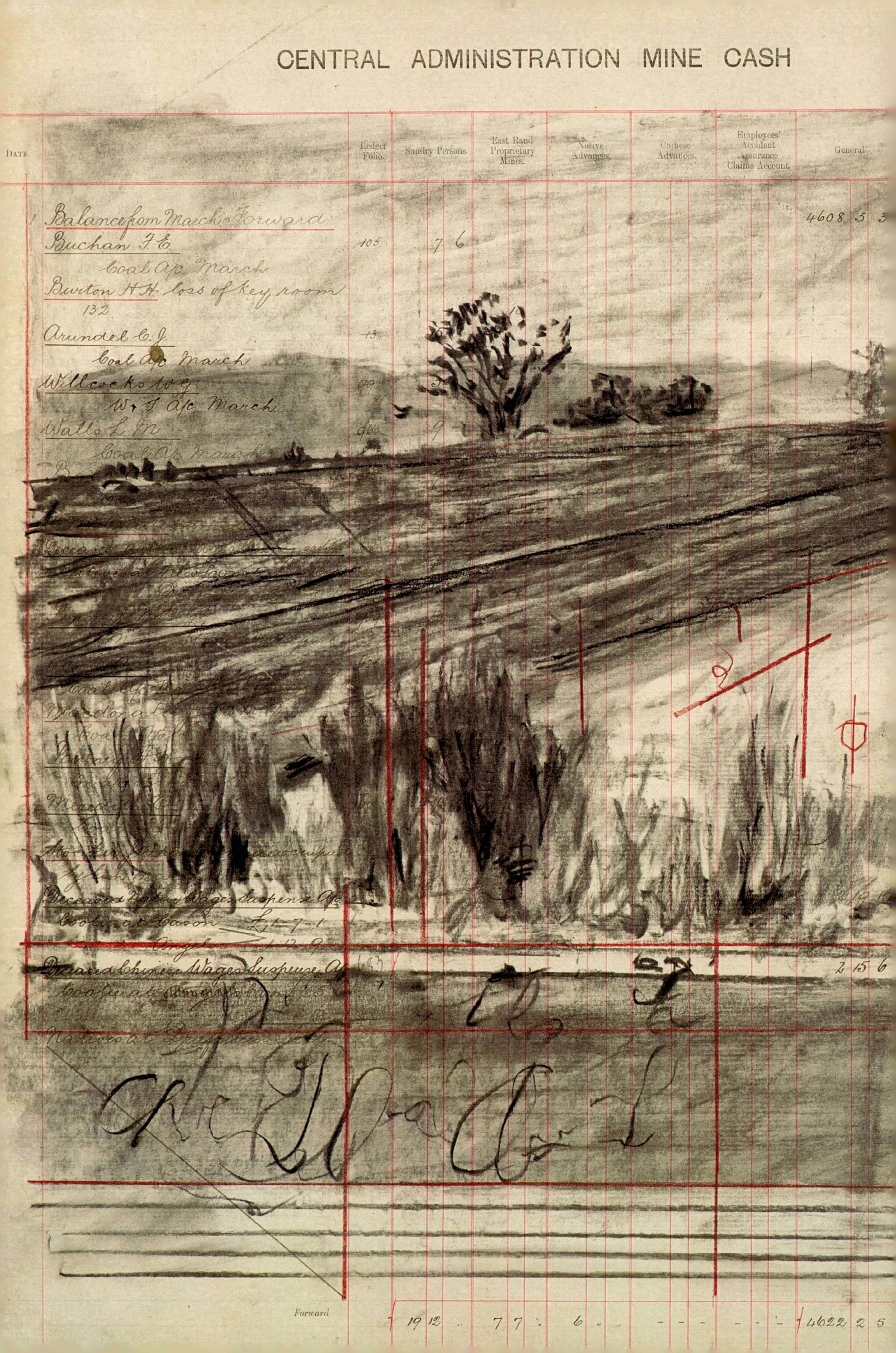
CENTRAL ADMINISTRATION MINE CASH
Date.
Ledger Folio.
Sundry Persons
East Rand Proprietary Mines.
Native Advances.
Chinese Advances.
Employees' Accident Assurance Claims Account.
General
Balance from March Forward
Buchan F.E.
Coal A/c March
Burton H.H. loss of key room
Arundel C.J.
Coal A/c March
Willcocks W.G.
Walls L.M.
Forward

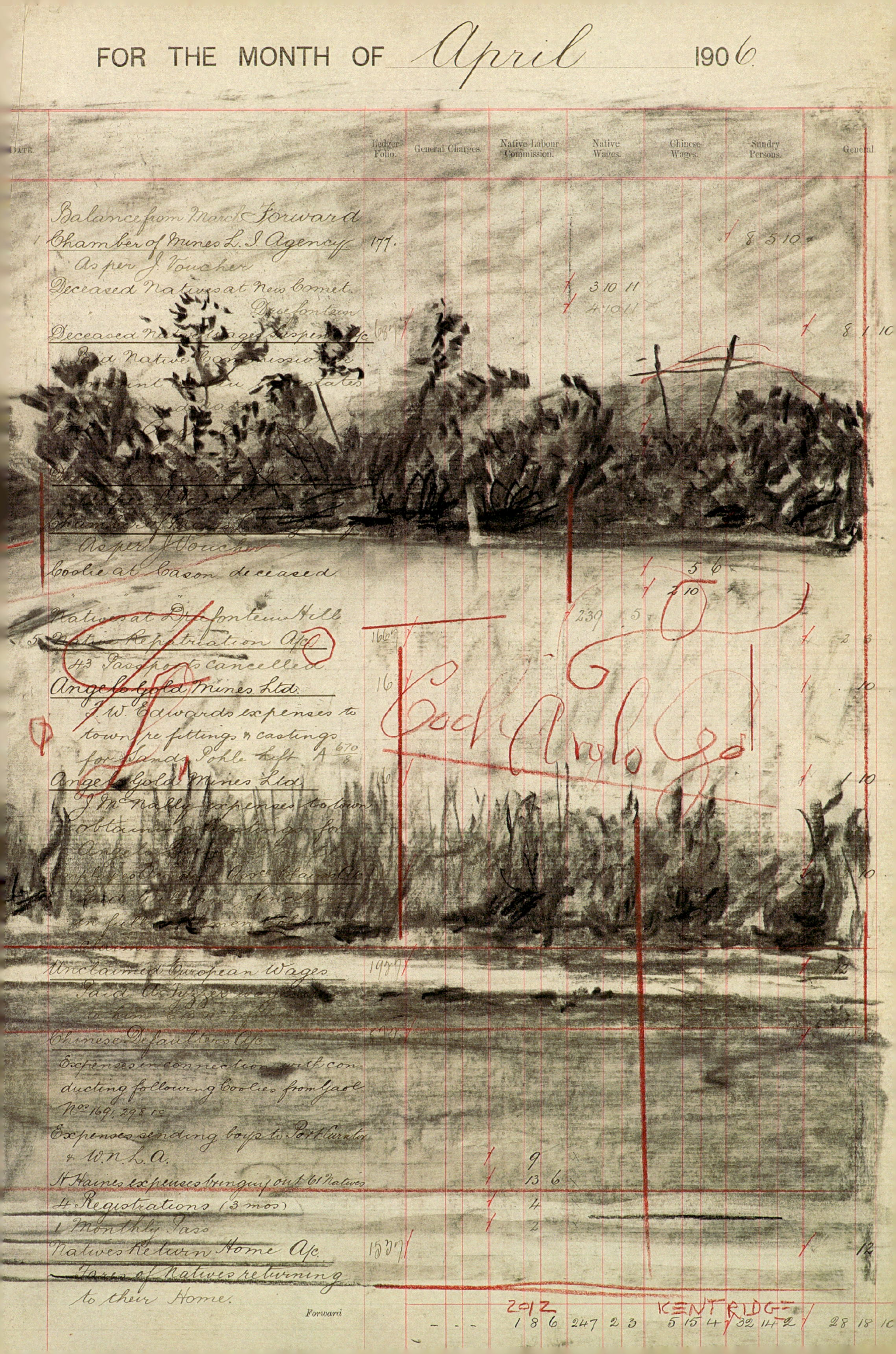

FOR THE MONTH OF April 1906
Ledger Folio.
General Charges.
Native Labour Commission.
Native Wages.
Chinese Wages.
Sundry Persons.
General.
Balance from March Forward
Chamber of Mines L. I. Agency
As per J. Voucher
Deceased Natives at New Comet
Coolie at Cason deceased
43 Passports cancelled
Angelo Gold Mines Ltd.
J. W. Edwards expenses to town re fittings & castings
Angelo Gold Mines Ltd
Unclaimed European Wages
Chinese Defaulters A/c.
Expenses in connection with conducting following Coolies from Gaol
Expenses sending boys to Port Curator & W.N.L.A.
H Haines expenses bringing out 61 Natives
4 Registrations (3 mos)
1 Monthly Pass
Natives Return Home A/c.
Fares of Natives returning to their Home.
Forward
Cock Anglo Go
KENTRIDGE

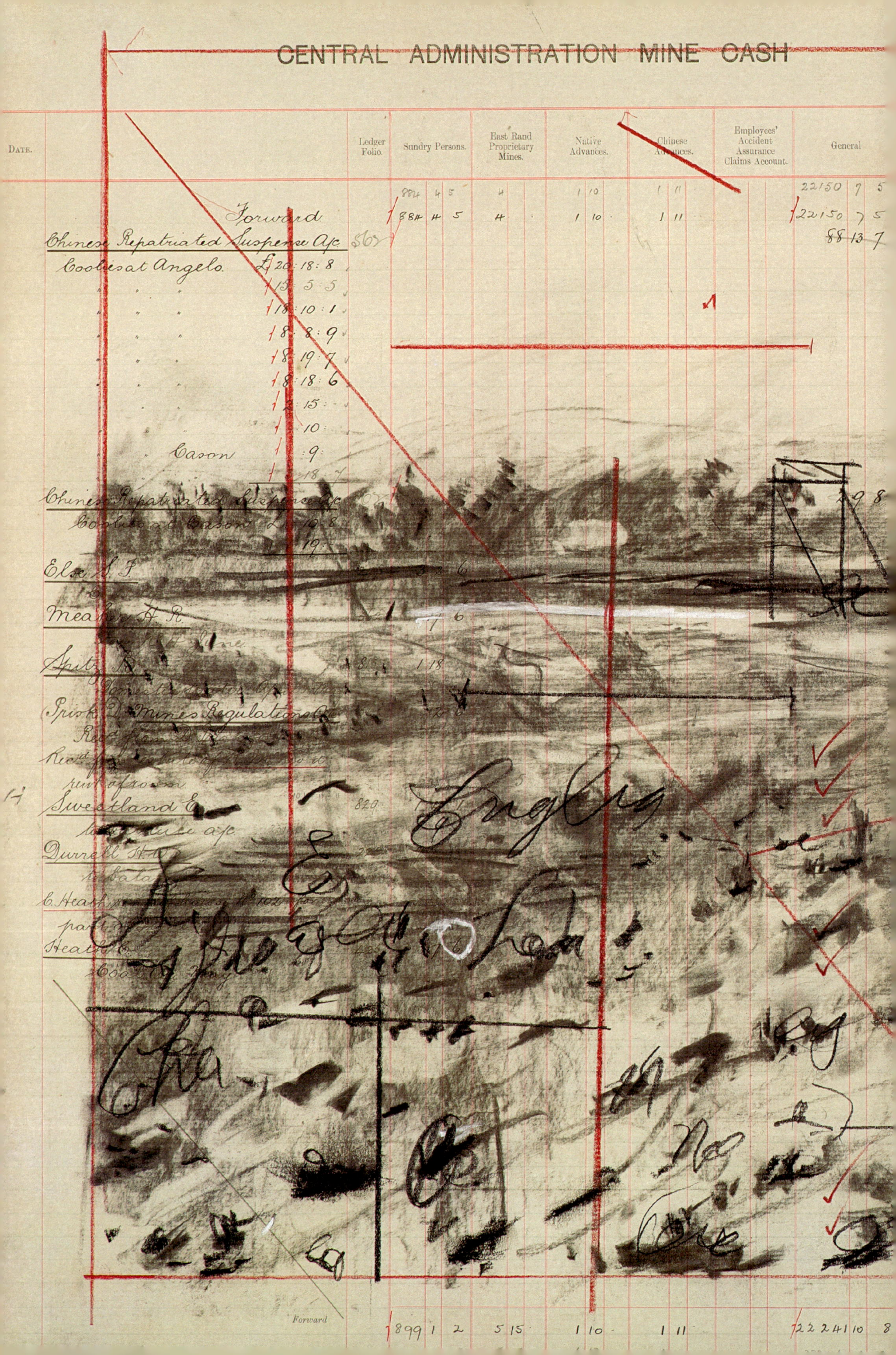
CENTRAL ADMINISTRATION MINE CASH
Date.
Ledger Folio.
Sundry Persons.
East Rand Proprietary Mines.
Native Advances.
Chinese Advances.
Employees' Accident Assurance Claims Account.
General.
Forward

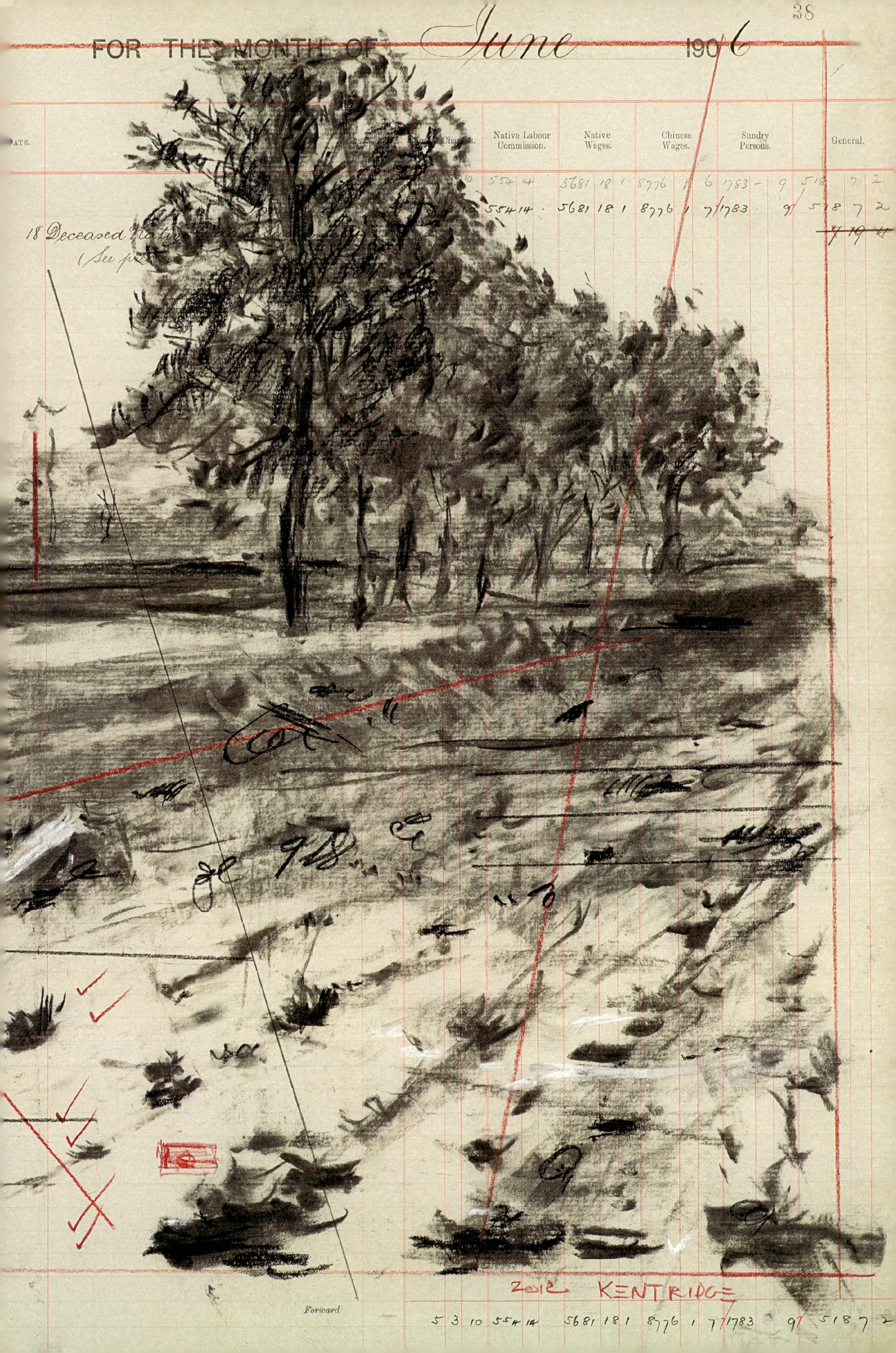
38
FOR THE MONTH OF June 1906
Native Labour Commission.
Native Wages.
Chinese Wages.
Sundry Persons.
General.
18 Deceased Nati
Forward
2012 KENTRIDGE

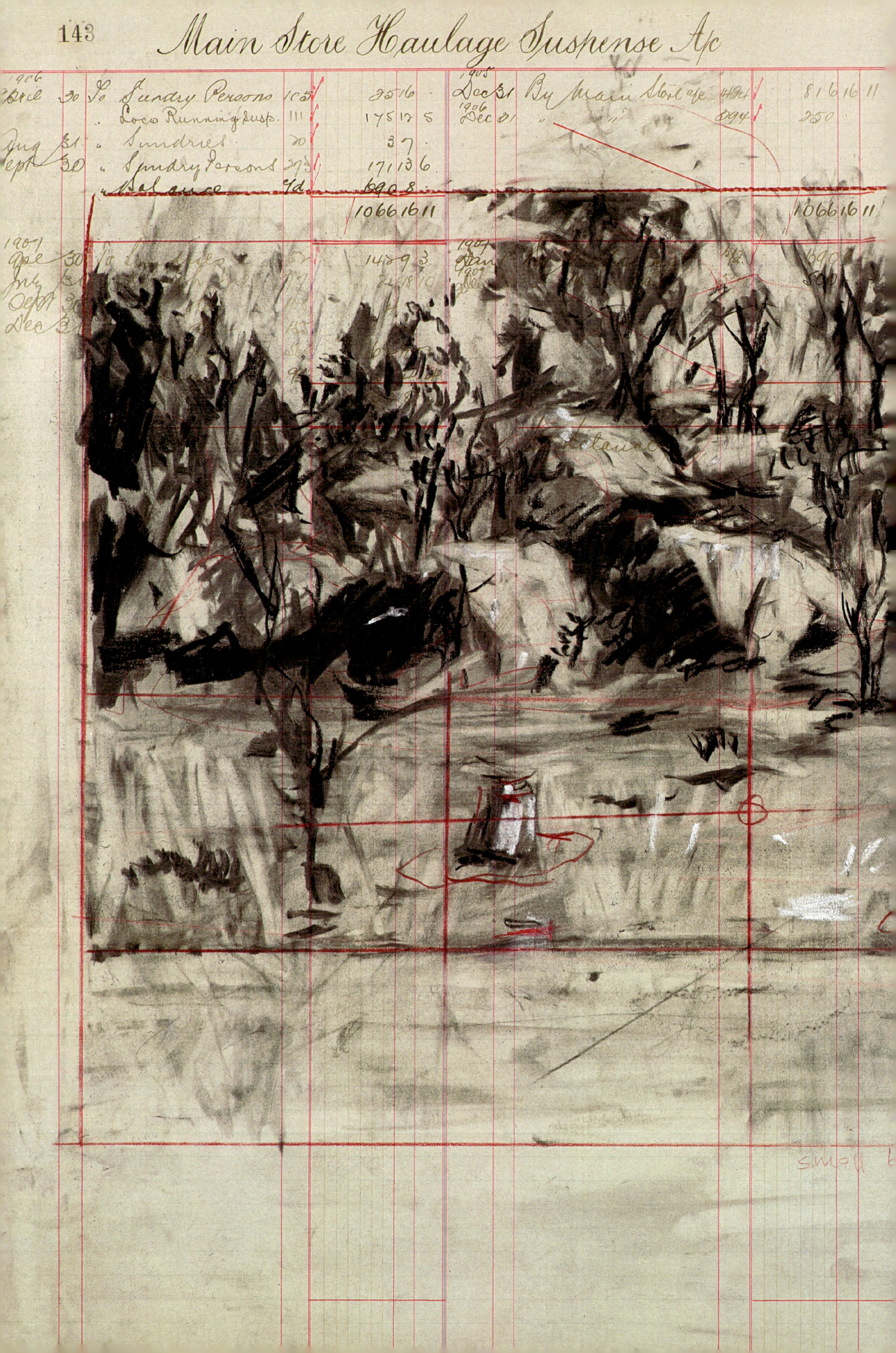
143
Main Store Haulage Suspense A/c

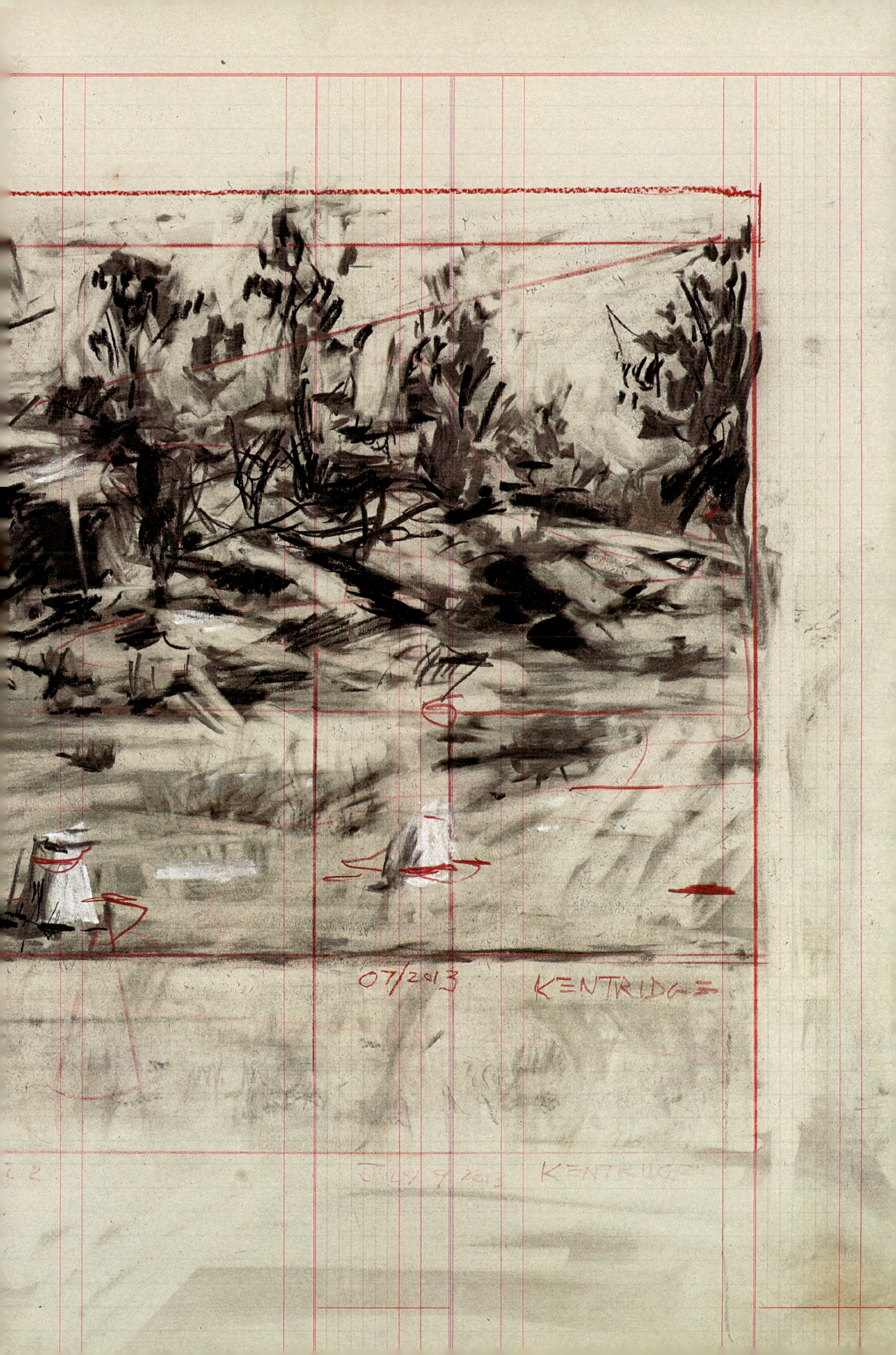
07/2013
KENTRIDGE

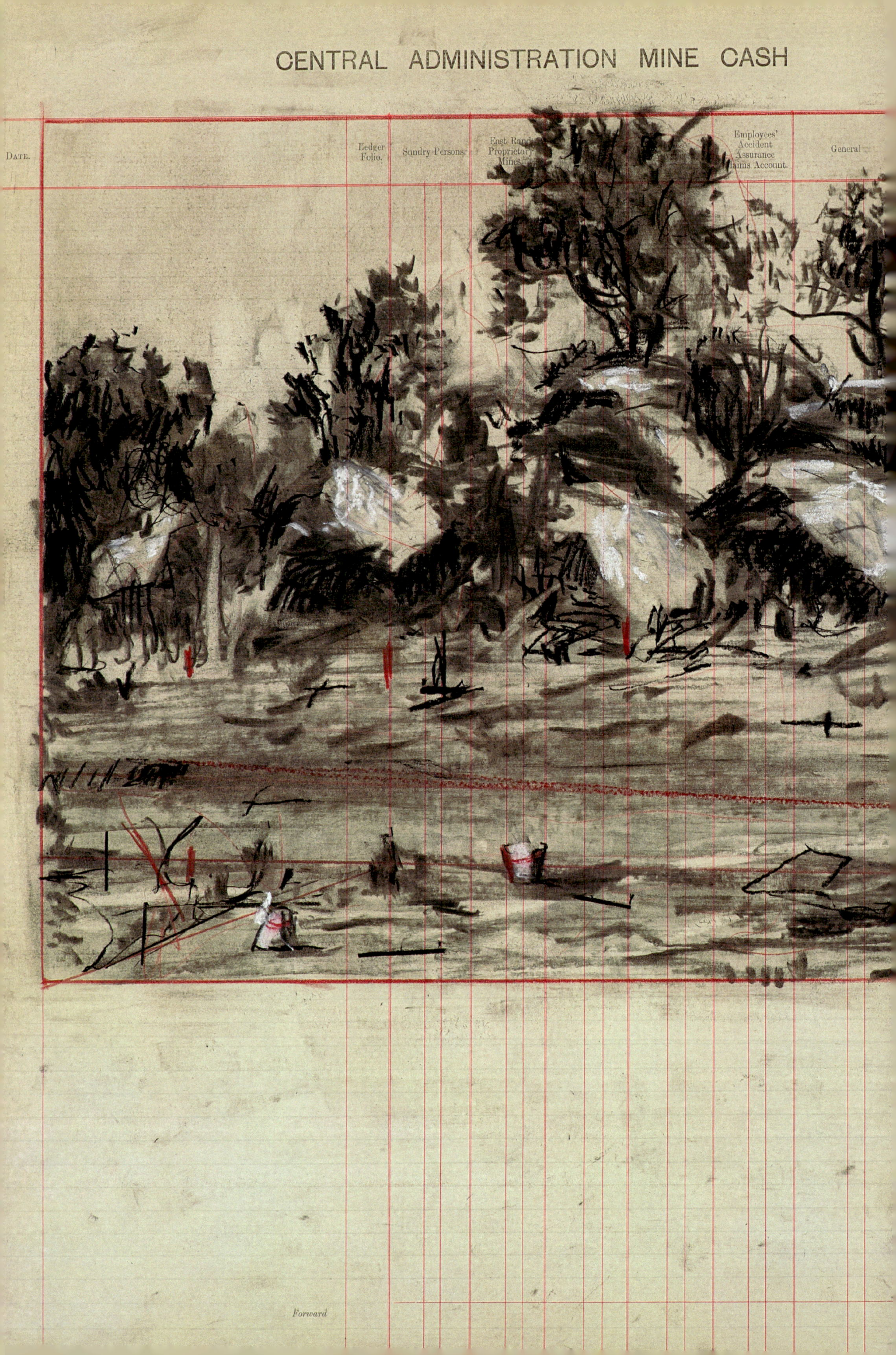
CENTRAL ADMINISTRATION MINE CASH
DATE.
Ledger Folio.
Sundry Persons.
East Rand Proprietary Mines
Employees' Accident Assurance
General
Forward

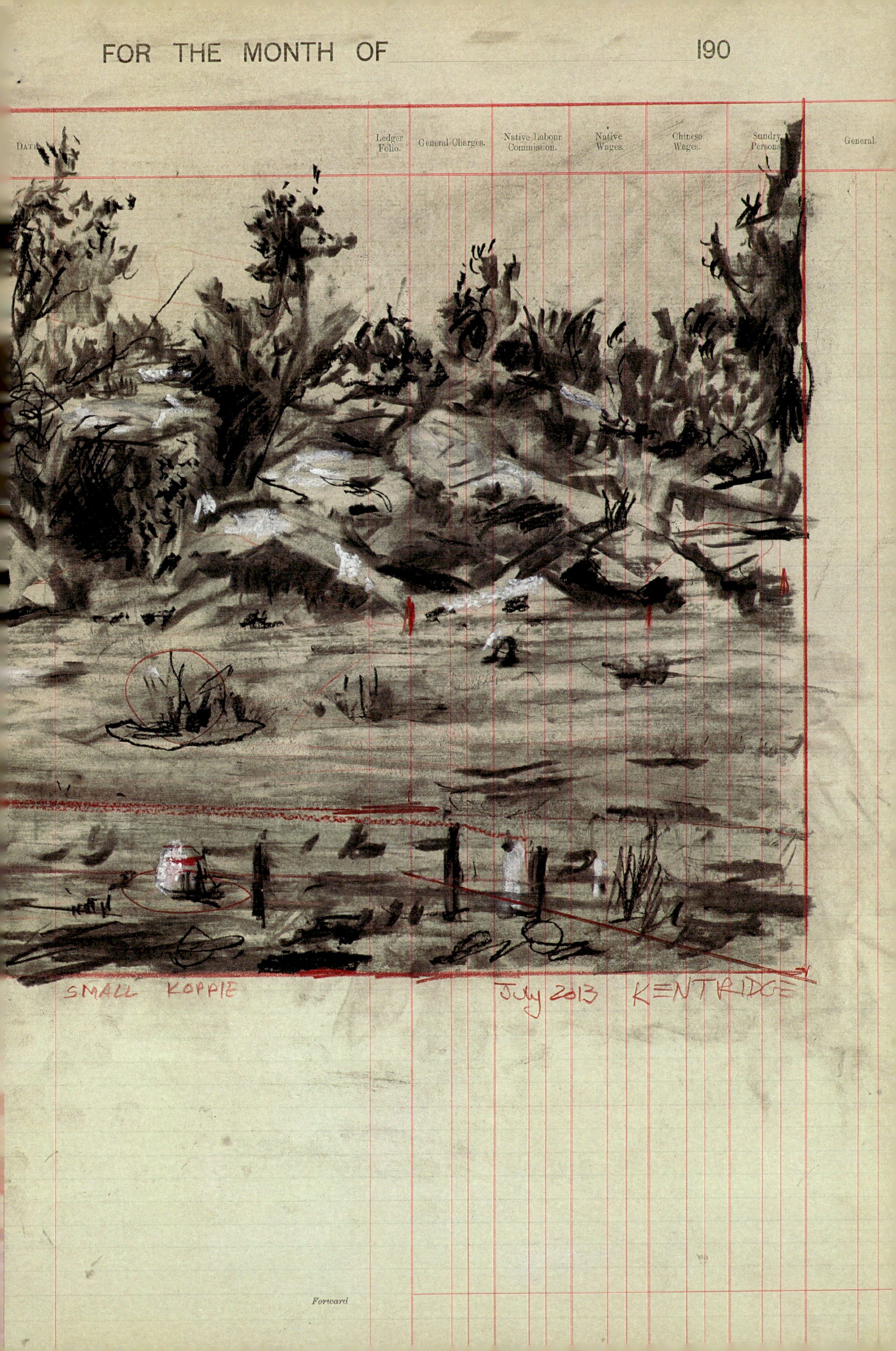
FOR THE MONTH OF
190
Date
Ledger Folio.
General Charges.
Native Labour Commission.
Native Wages.
Chinese Wages.
Sundry Persons.
General.
SMALL KOPPIE
July 2013
KENTRIDGE
Forward

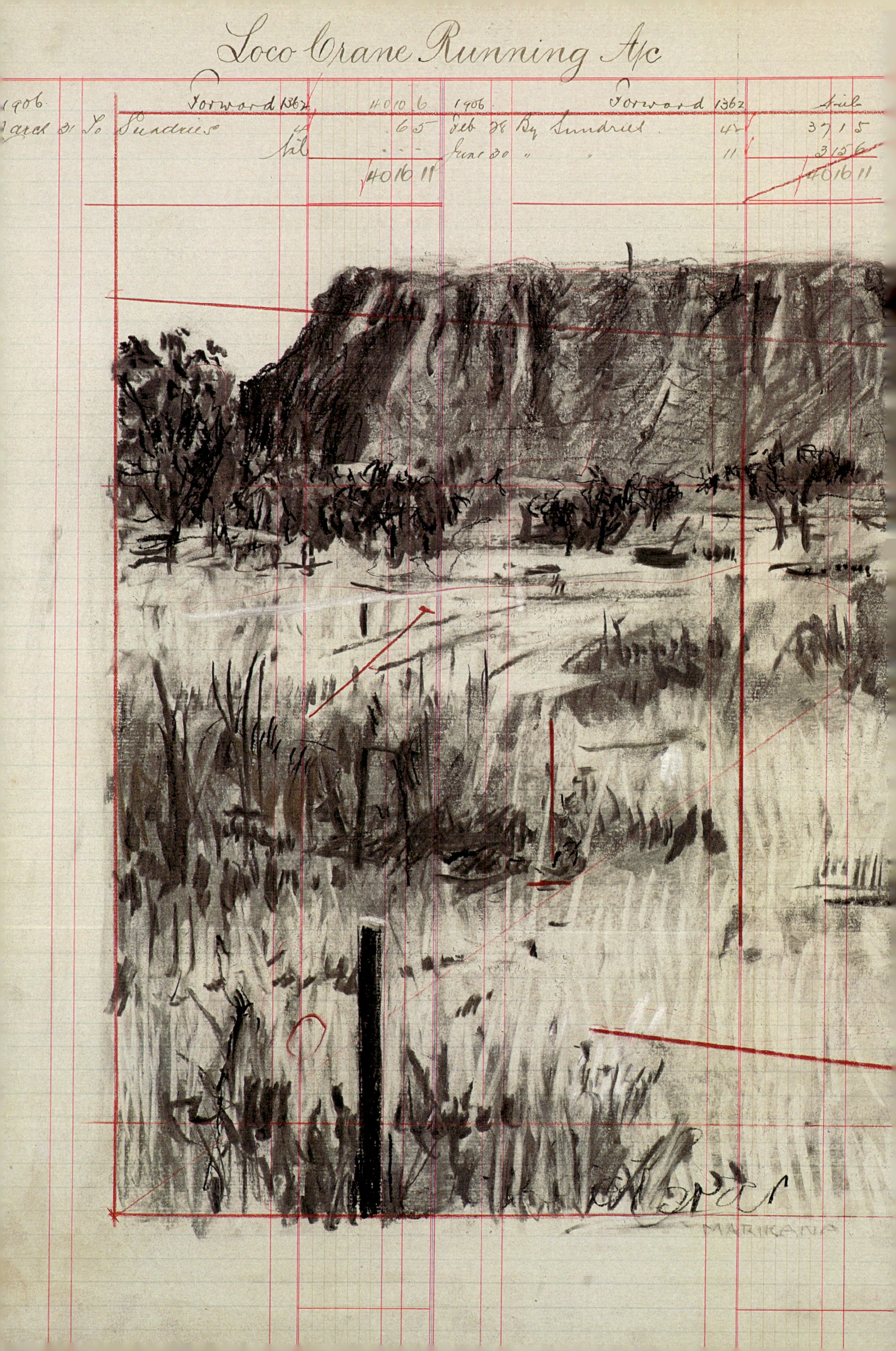
Loco Crane Running A/c
1906
Forward 1362
40 10 6
March 31 To Sundries
Feb 28 By Sundries
June 30
40 16 11
MARIKANA

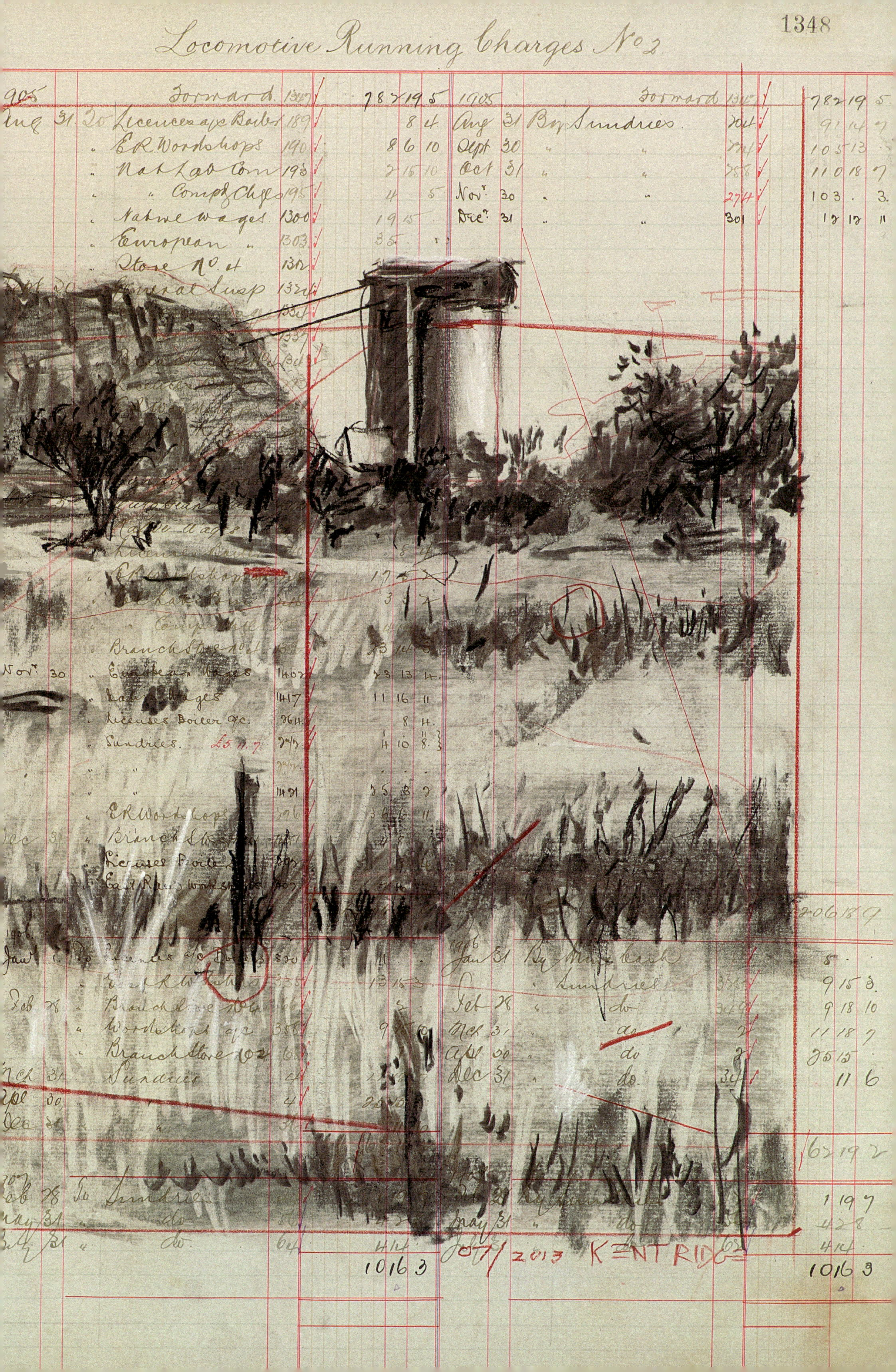
1348
Locomotive Running Charges No 2
KENTRIDGE

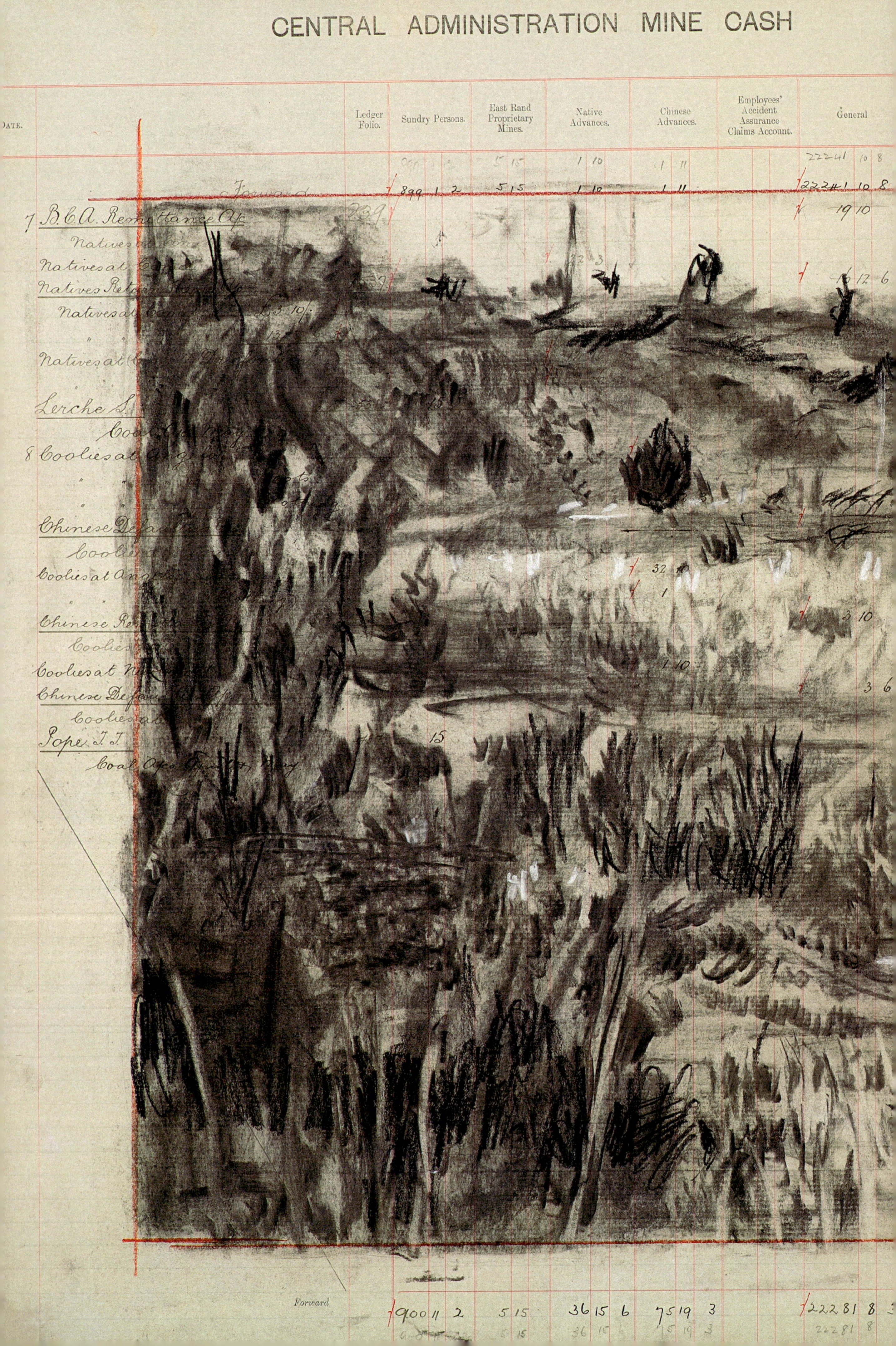

CENTRAL ADMINISTRATION MINE CASH
DATE.
Ledger Folio.
Sundry Persons.
East Rand Proprietary Mines.
Native Advances.
Chinese Advances.
Employees' Accident Assurance Claims Account.
General
B.C.A. Remittance A/c
Natives at
Natives Ret
Lerche S.
Coolies at
Chinese Def
Coolies at
Chinese Re
Coolies at
Chinese Def
Pope. T. T.
Forward

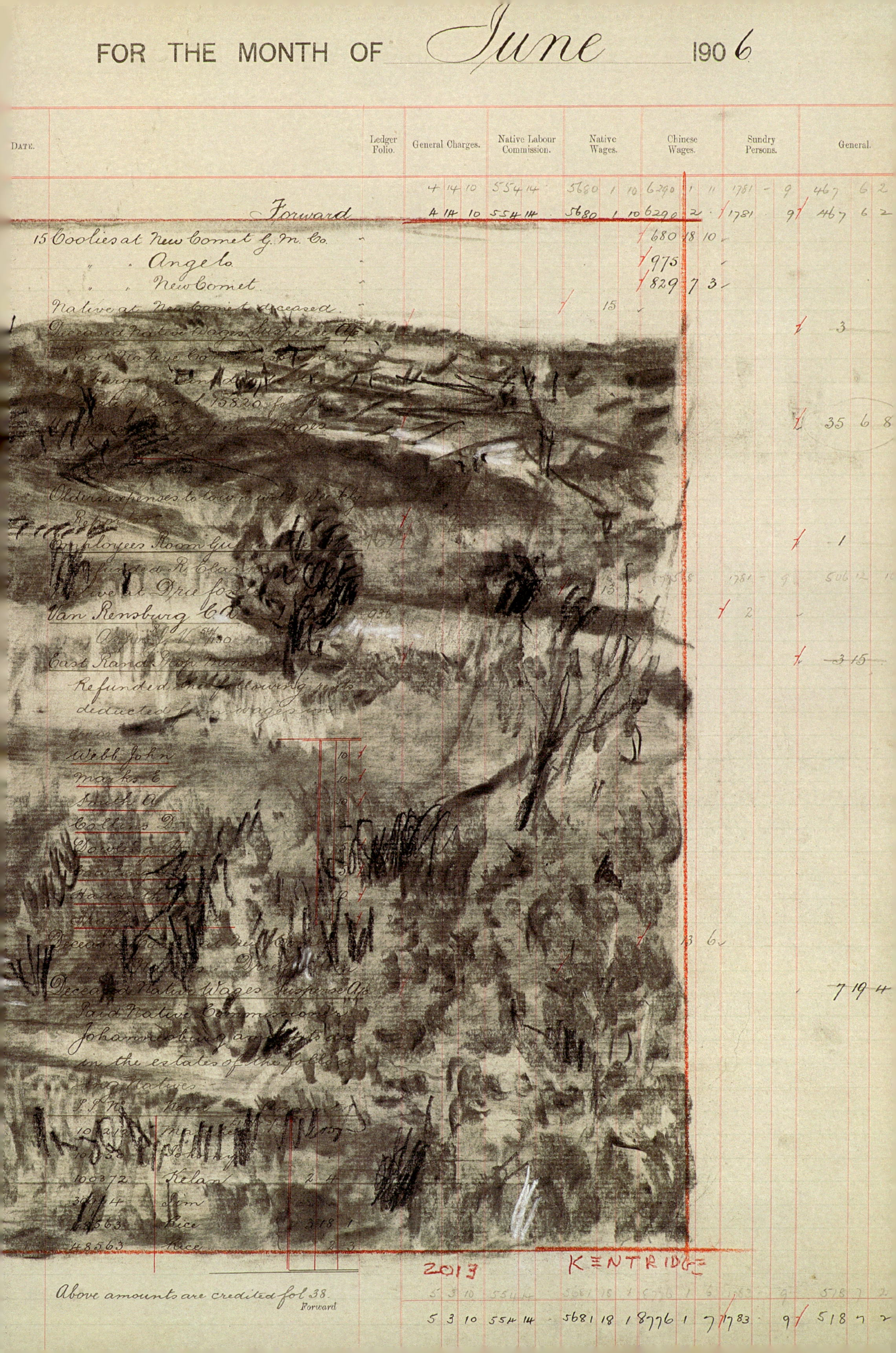
FOR THE MONTH OF June 1906
Date.
Ledger Folio.
General Charges.
Native Labour Commission.
Native Wages.
Chinese Wages.
Sundry Persons.
General.
Forward
15 Coolies at New Comet G. M. Co.
Angelo
New Comet
Van Rensburg C A
Webb John
Above amounts are credited fol 38
Forward
2013
KENTRIDGE

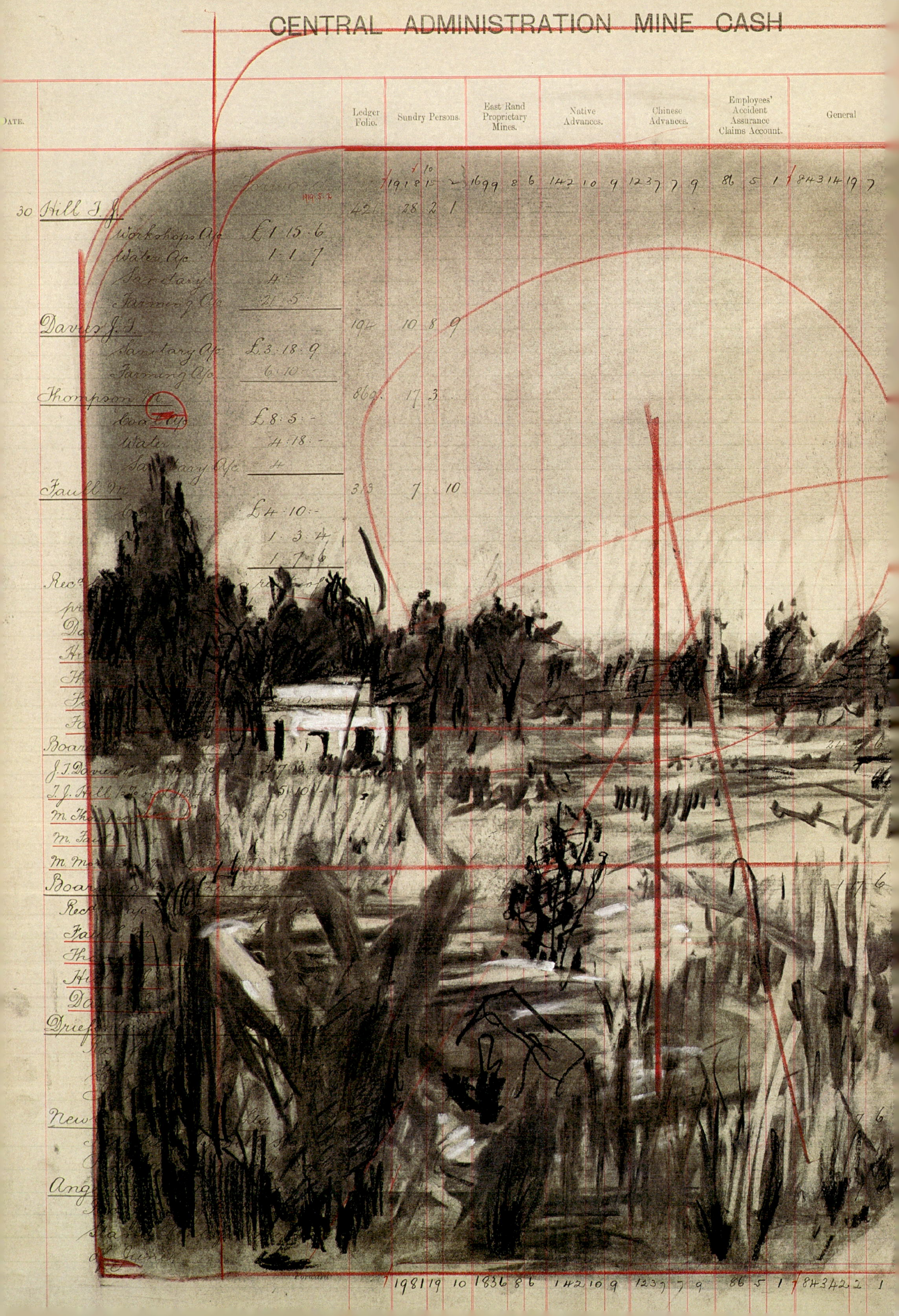
CENTRAL ADMINISTRATION MINE CASH
DATE.
Ledger Folio.
Sundry Persons.
East Rand Proprietary Mines.
Native Advances.
Chinese Advances.
Employees' Accident Assurance Claims Account.
General.

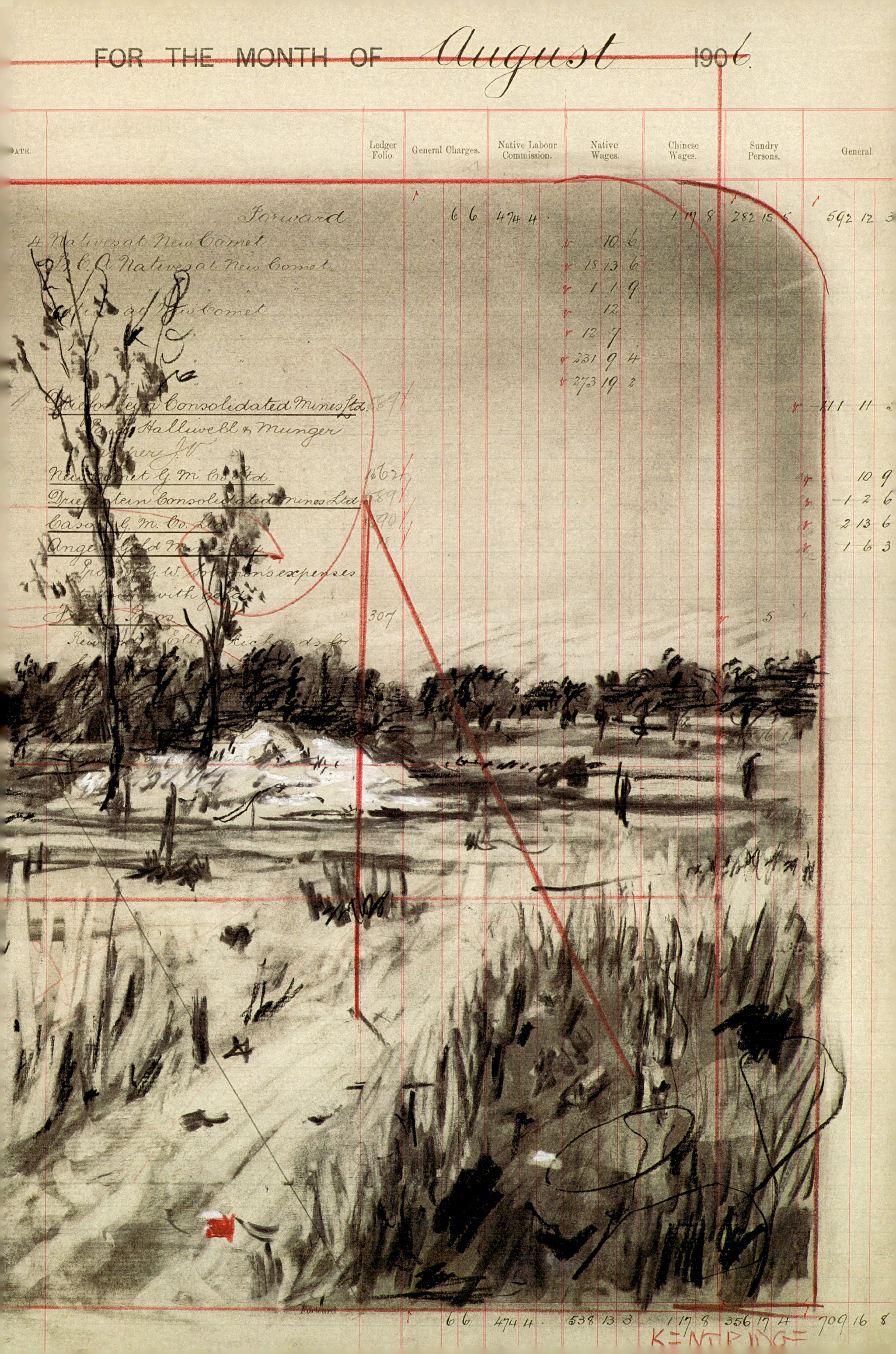
FOR THE MONTH OF August 1906.
Ledger Folio.
General Charges.
Native Labour Commission.
Native Wages.
Chinese Wages.
Sundry Persons.
General.
Forward
KENTRIDGE

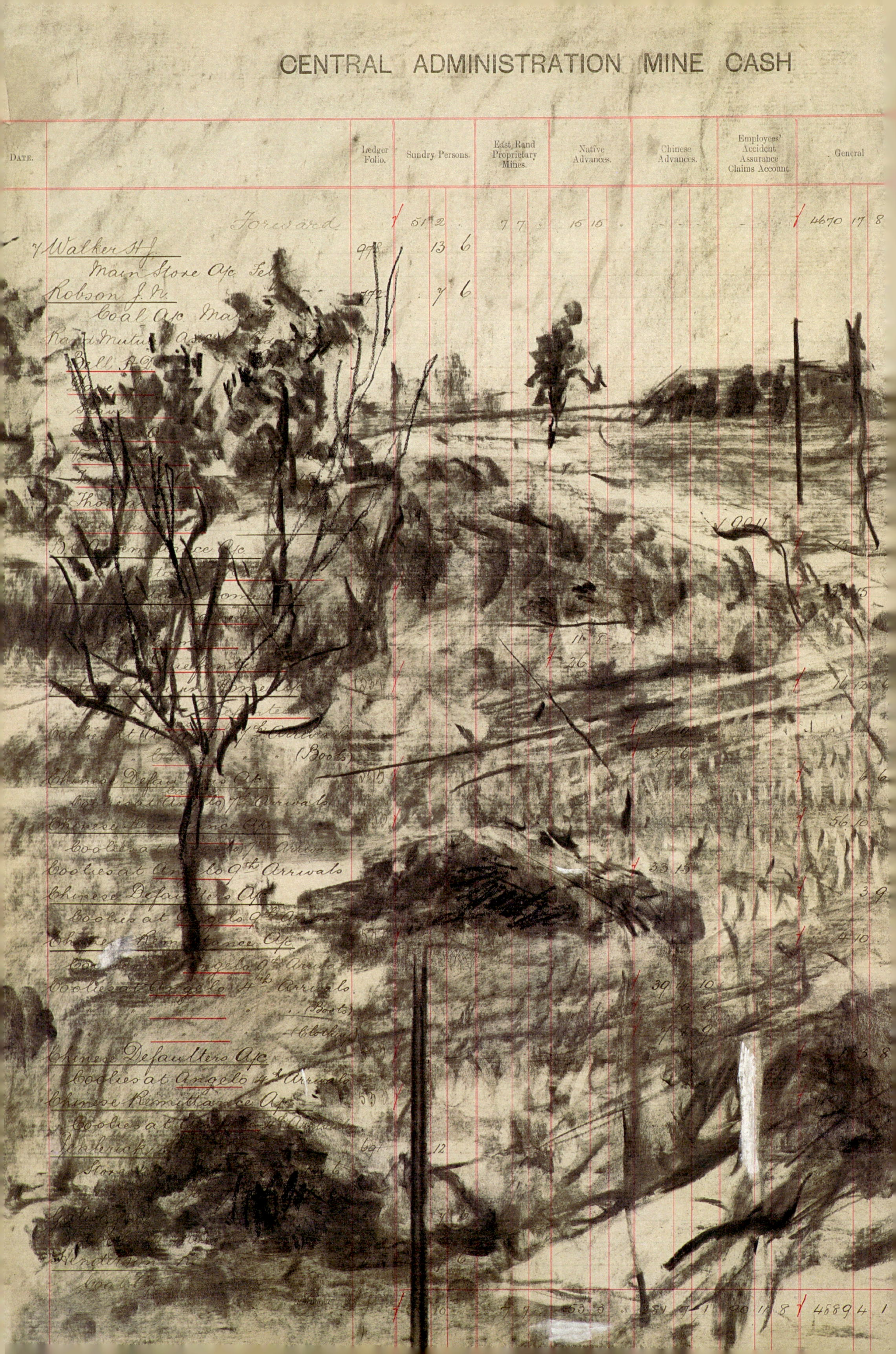
CENTRAL ADMINISTRATION MINE CASH
Date.
Ledger Folio.
Sundry Persons.
East Rand Proprietary Mines.
Native Advances.
Chinese Advances.
Employees' Accident Assurance Claims Account.
General
Forward

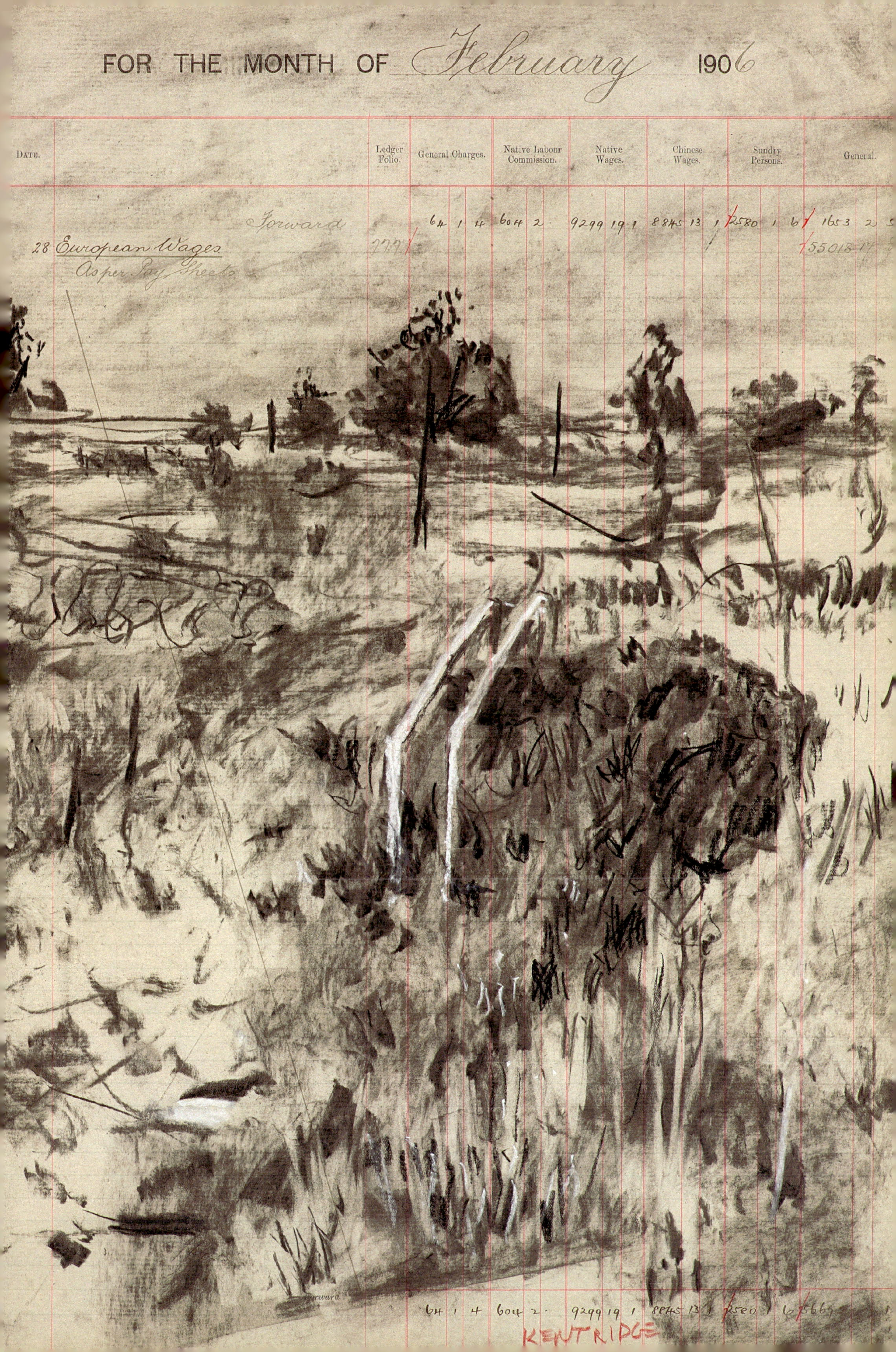
FOR THE MONTH OF February 1906
Date.
Ledger Folio.
General Charges.
Native Labour Commission.
Native Wages.
Chinese Wages.
Sundry Persons.
General.
Forward
28 European Wages
As per Pay Sheets
KENTRIDGE

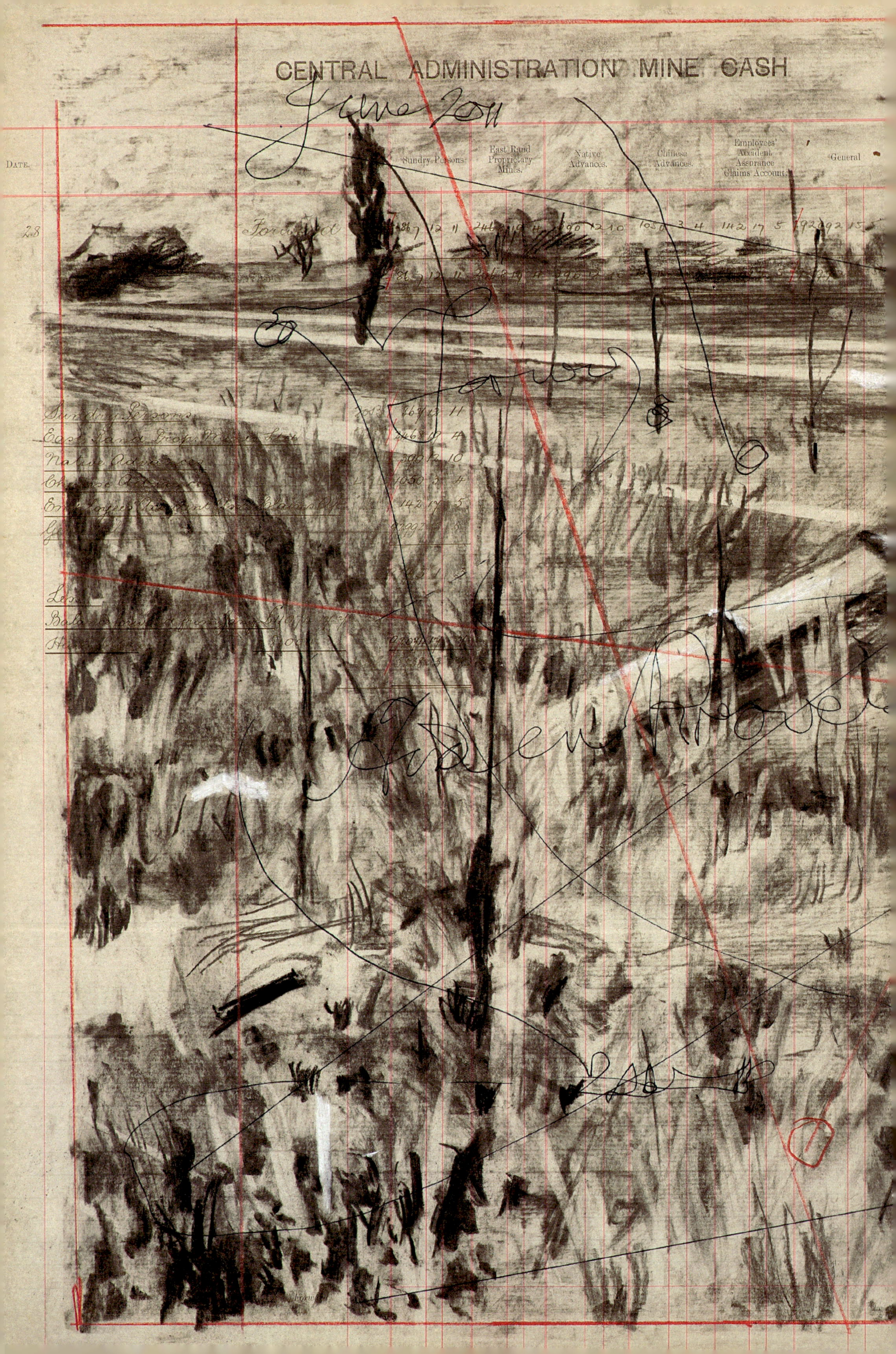
CENTRAL ADMINISTRATION MINE CASH
Date.
Sundry Persons.
East Rand Proprietary Mines.
Native Advances.
Chinese Advances.
Employees' Accident Assurance Claims Account.
General

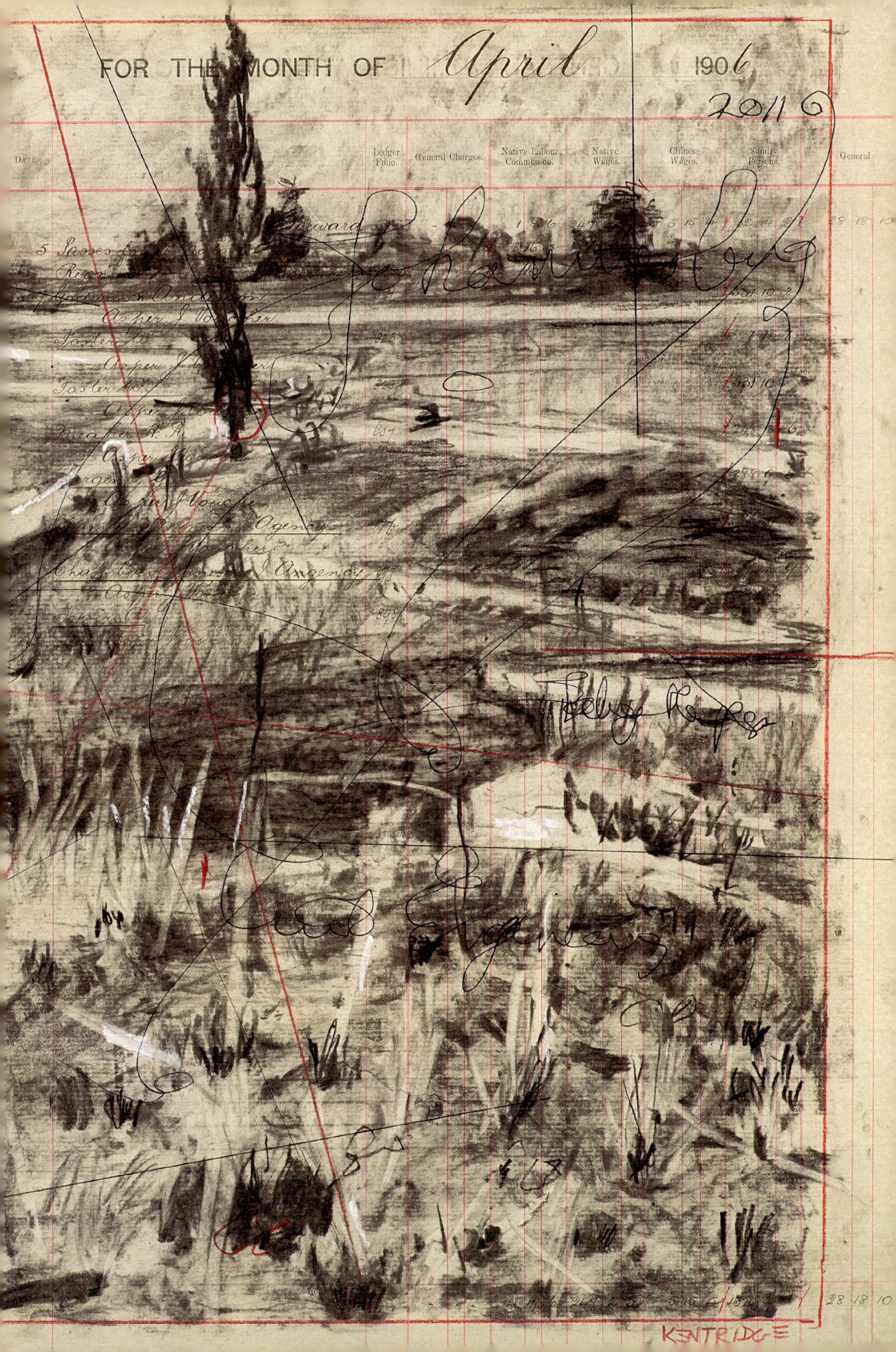
FOR THE MONTH OF April 1906
2011
Ledger Folio.
General Charges.
Native Labour Commission.
Native Wages.
Chinese Wages.
Sundry Persons.
General.
KENTRIDGE

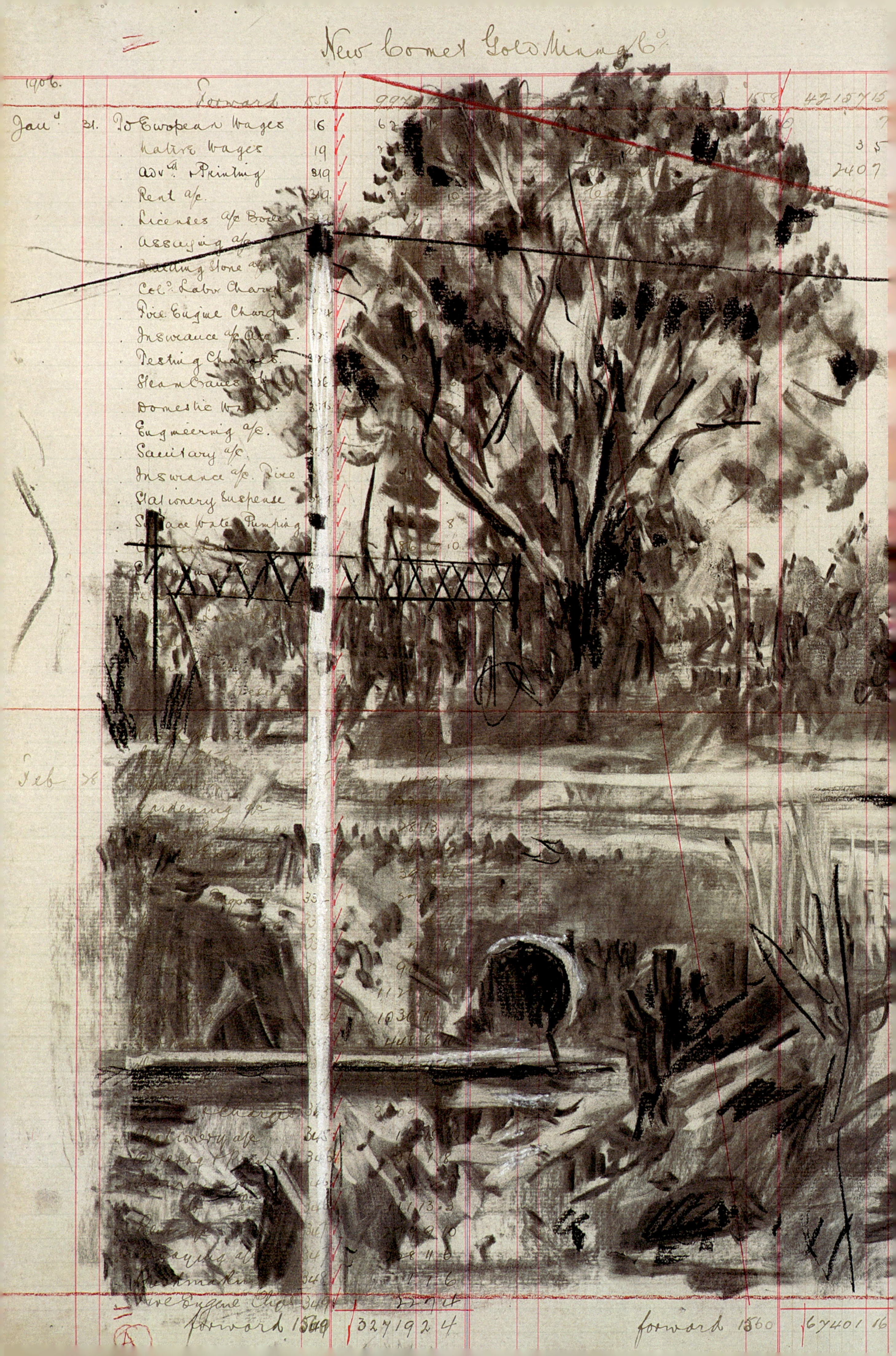
New Comet Gold Mining Co
1906.
Forward
Jan 31. To European Wages
Native Wages
Advt. Printing
Rent a/c
Licenses
Assaying a/c
Col. Labor Charges
Fire Engine Charges
Insurance a/c
Testing Charges
Steam Cranes
Domestic
Engineering a/c
Sanitary a/c
Insurance a/c Fire
Stationery Suspense
Surface Water Pumping
Feb
Gardening a/c
forward 1549
3271924
forward 1560
67401 16

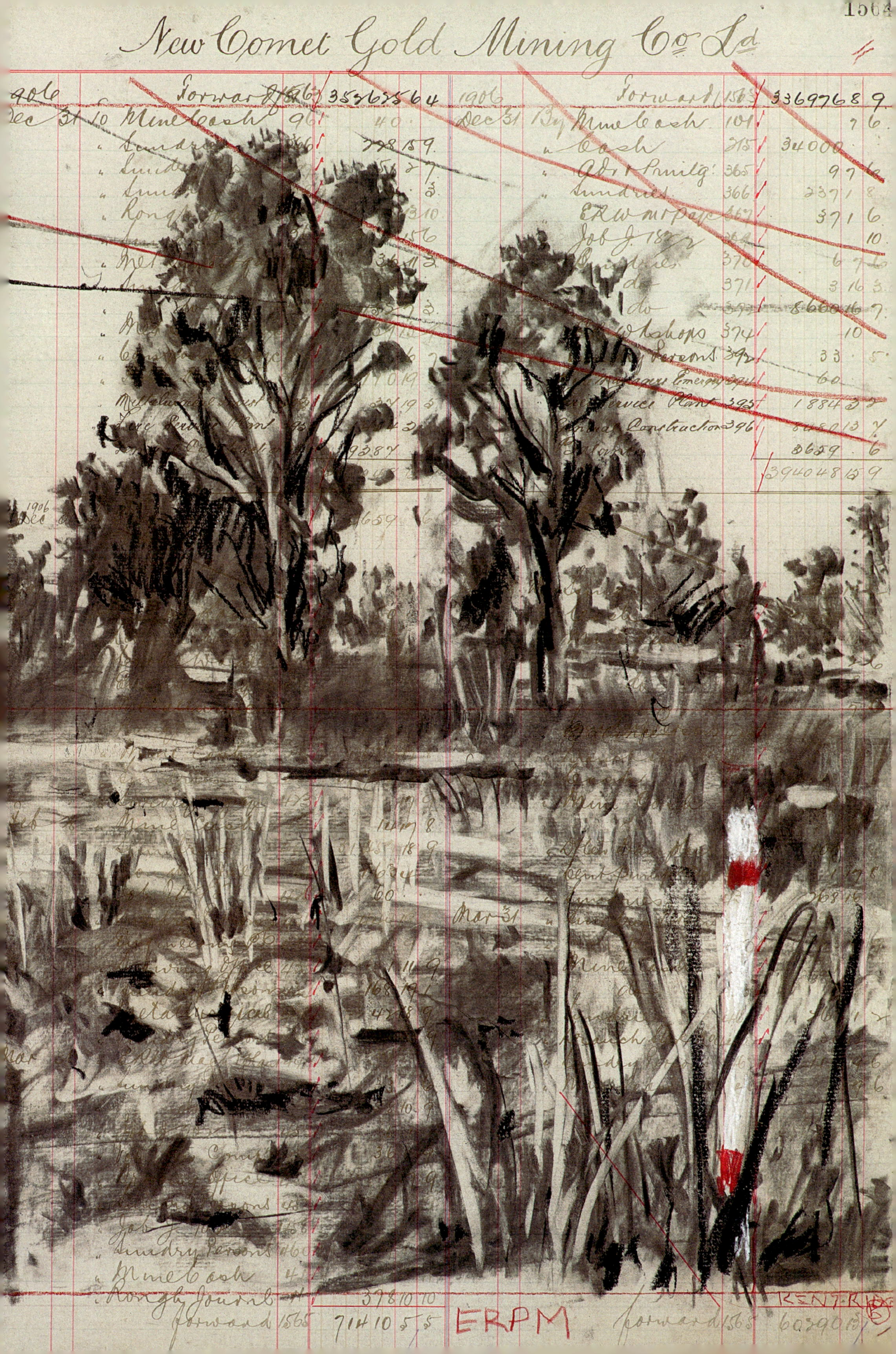

New Comet Gold Mining Co Ld
ERPM

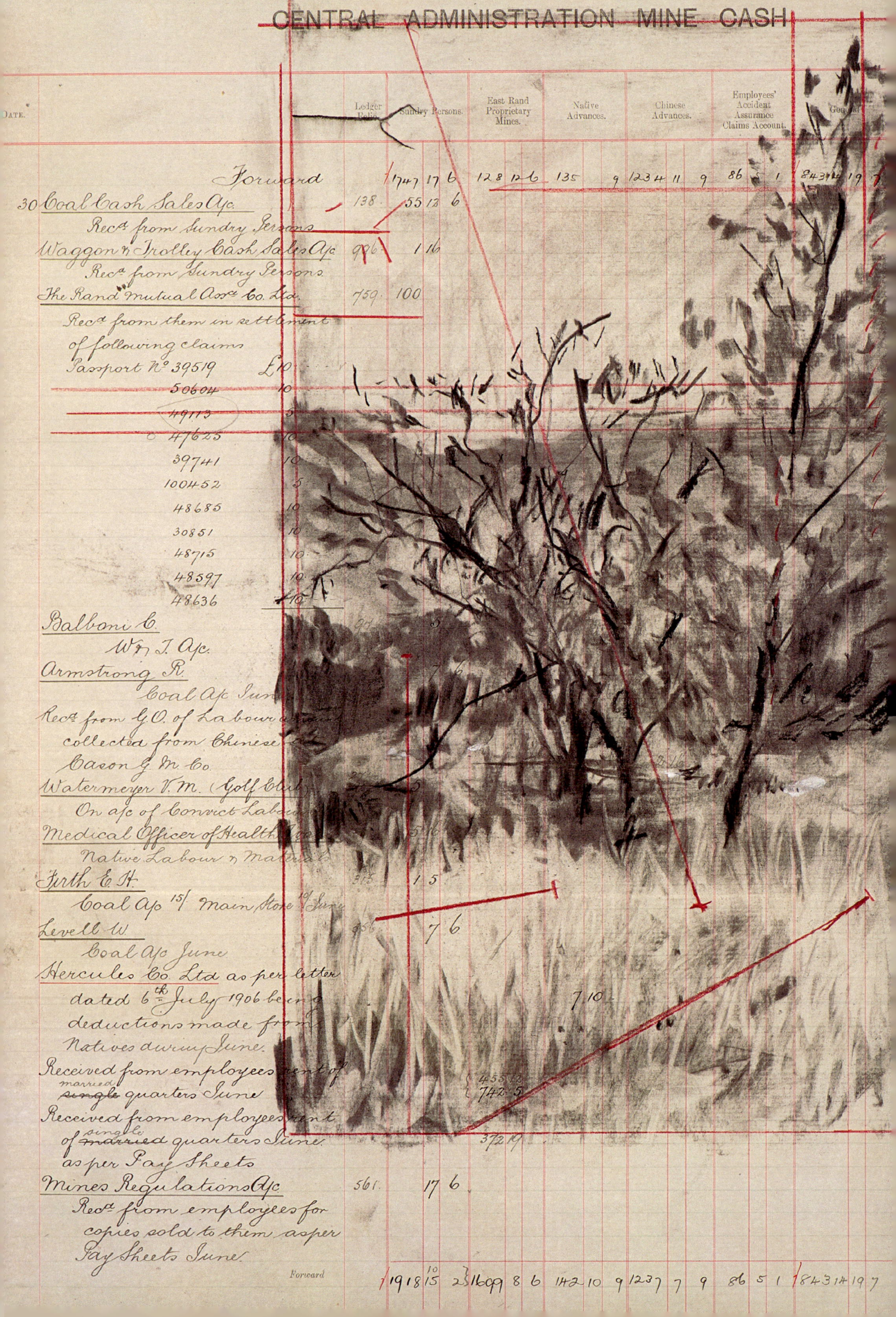

CENTRAL ADMINISTRATION MINE CASH
Date.
Ledger Folio
Sundry Persons.
East Rand Proprietary Mines.
Native Advances.
Chinese Advances.
Employees' Accident Assurance Claims Account.
Forward
30 Coal Cash Sales A/c
Recd from Sundry Persons
Waggon & Trolley Cash Sales A/c
Recd from Sundry Persons
The Rand Mutual Assce Co. Ltd.
Recd from them in settlement
of following claims
Passport No 39519
50604
49113
47623
39741
100452
48685
30851
48715
48597
48636
Balboni C.
W & T. A/c
Armstrong R.
Coal A/c June
Recd from G.O. of Labour
collected from Chinese
Cason G. M. Co.
Watermeyer V. M. Golf Club
On a/c of Convict Labour
Medical Officer of Health
Native Labour & Material
Firth E H.
Coal A/c 15/ Main Store 10/ June
Levell W.
Coal A/c June
Hercules Co. Ltd as per letter
dated 6th July 1906 being
deductions made from
Natives during June.
Received from employees rent of
married quarters June
Received from employees rent
of single quarters June
as per Pay Sheets
Mines Regulations A/c
Recd from employees for
copies sold to them as per
Pay Sheets June
Forward

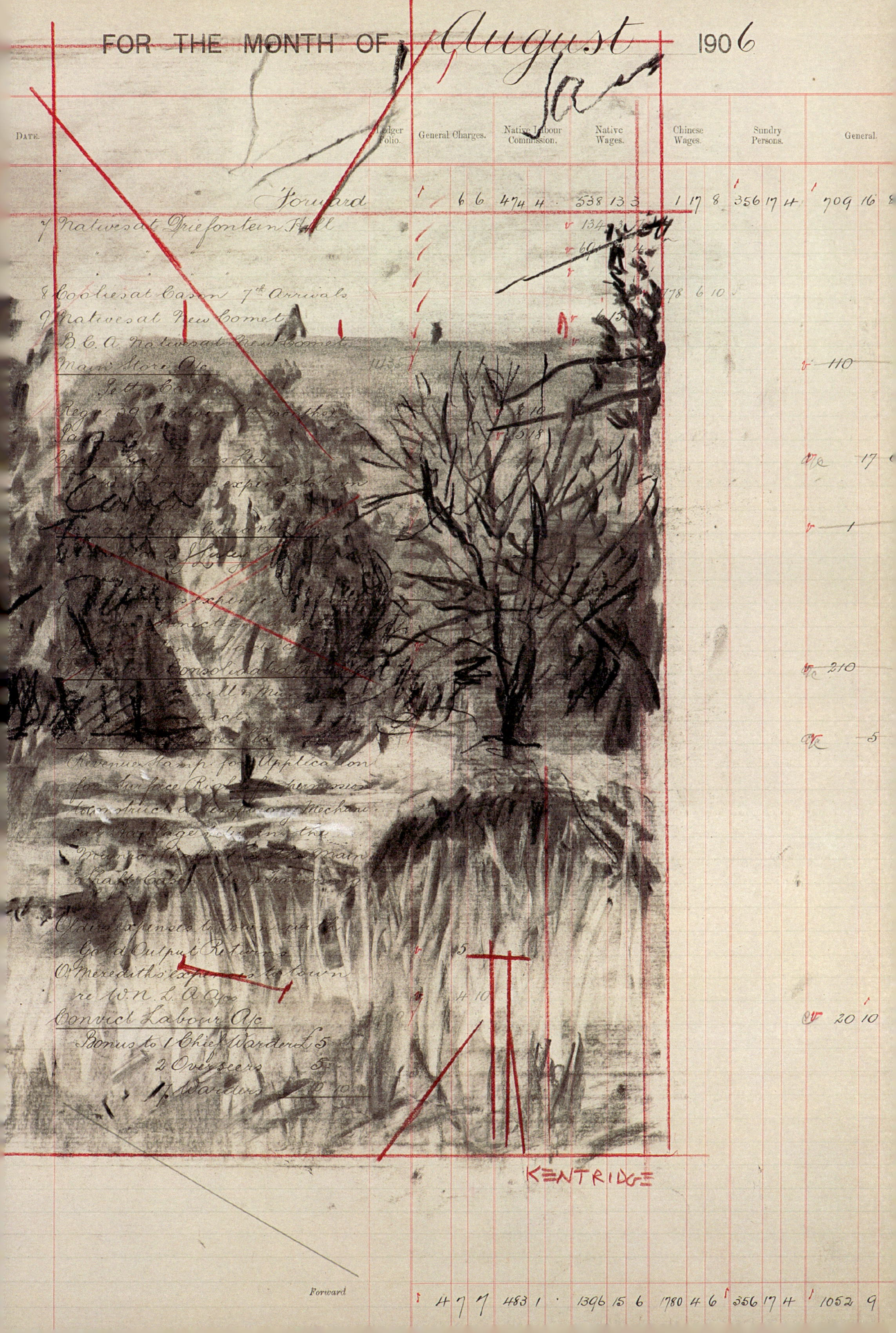
FOR THE MONTH OF August 1906
Date.
Ledger Folio.
General Charges.
Native Labour Commission.
Native Wages.
Chinese Wages.
Sundry Persons.
General.
Forward
Convict Labour A/c
KENTRIDGE
Forward

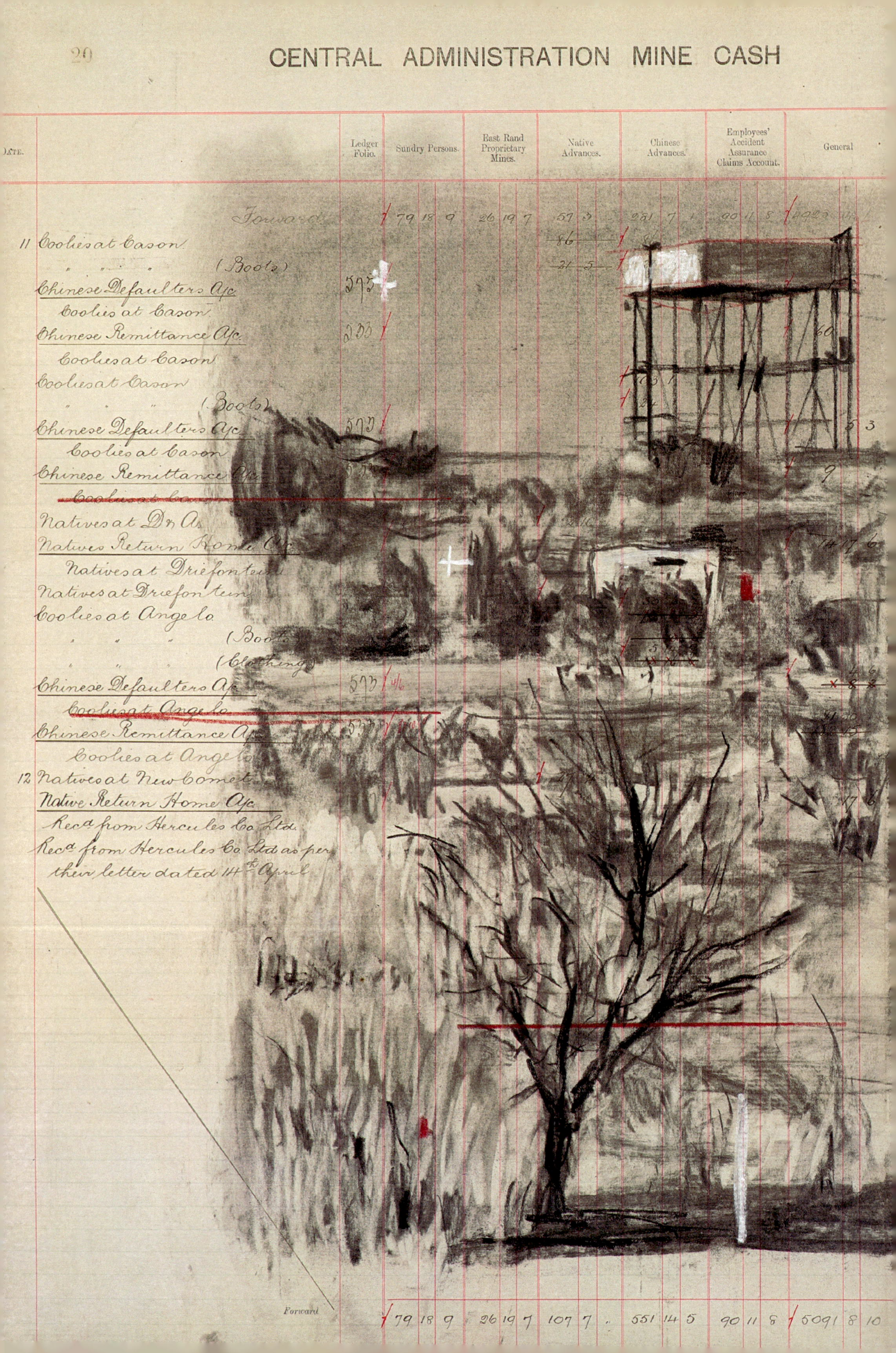

20
CENTRAL ADMINISTRATION MINE CASH
Date.
Ledger Folio.
Sundry Persons.
East Rand Proprietary Mines.
Native Advances.
Chinese Advances.
Employees' Accident Assurance Claims Account.
General
Forward
11 Coolies at Cason
(Boots)
Chinese Defaulters A/c
Coolies at Cason
Chinese Remittance A/c
Coolies at Cason
Coolies at Cason
(Boots)
Chinese Defaulters A/c
Coolies at Cason
Chinese Remittance A/c
Natives at D&A
Natives Return Home A/c
Natives at Driefontein
Natives at Driefontein
Coolies at Angelo
(Boots)
(Clothing)
Chinese Defaulters A/c
Coolies at Angelo
Chinese Remittance A/c
Coolies at Angelo
12 Natives at New Comet
Native Return Home A/c
Recd from Hercules Co Ltd
Recd from Hercules Co Ltd as per their letter dated 14th April
Forward
79 18 9
26 19 7
107 7 .
551 14 5
90 11 8
5091 8 10

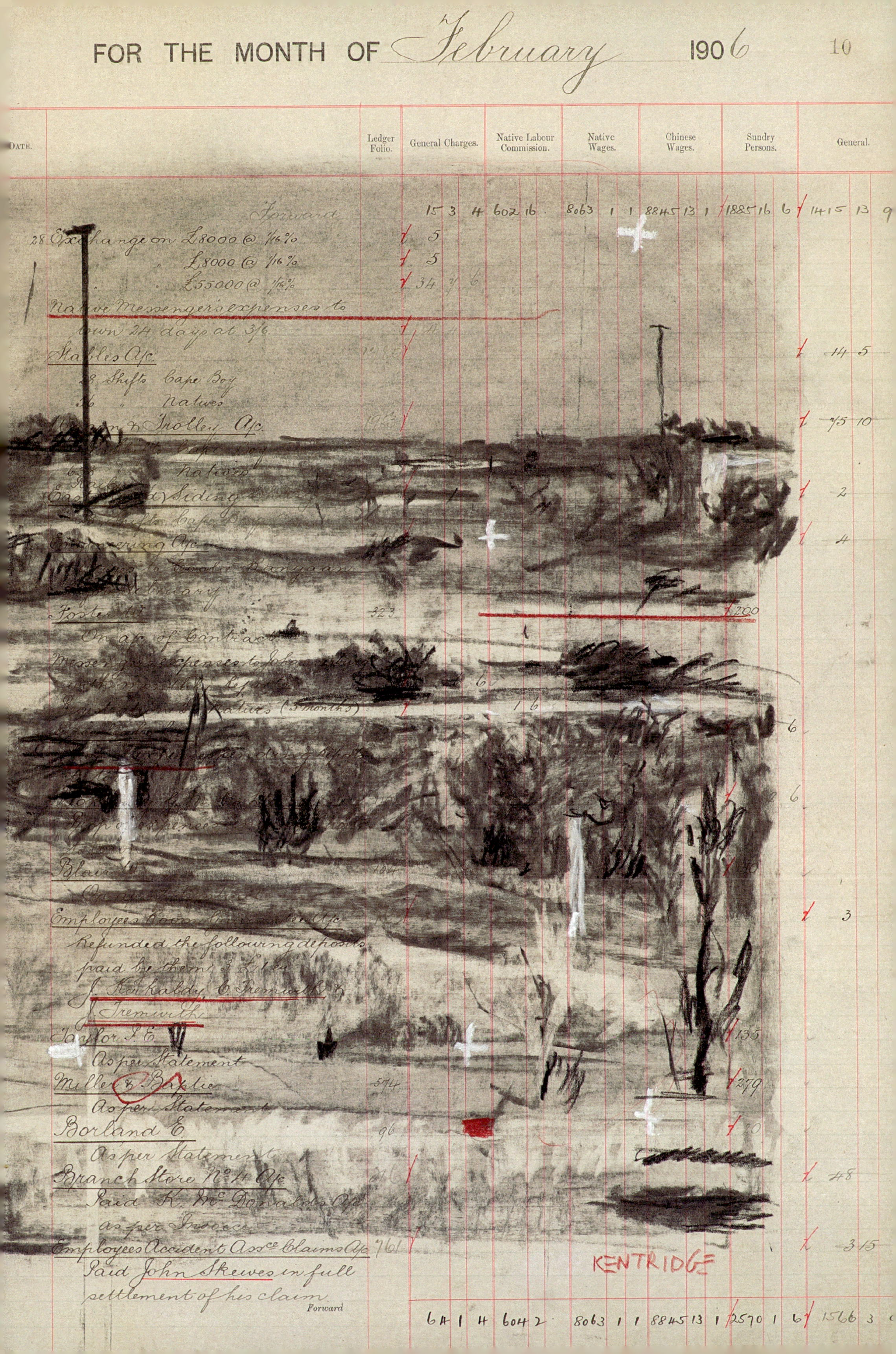

FOR THE MONTH OF February 1906
10
Date.
Ledger Folio.
General Charges.
Native Labour Commission.
Native Wages.
Chinese Wages.
Sundry Persons.
General.
Forward
Exchange on £8000 @ 1/16%
£8000 @ 1/16%
£55000 @ 1/16%
Native Messengers expenses to
town 24 days at 3/6
Stables A/c
Employees Accident Ass. Claims A/c
Paid John Skewes in full
settlement of his claim
Forward
KENTRIDGE

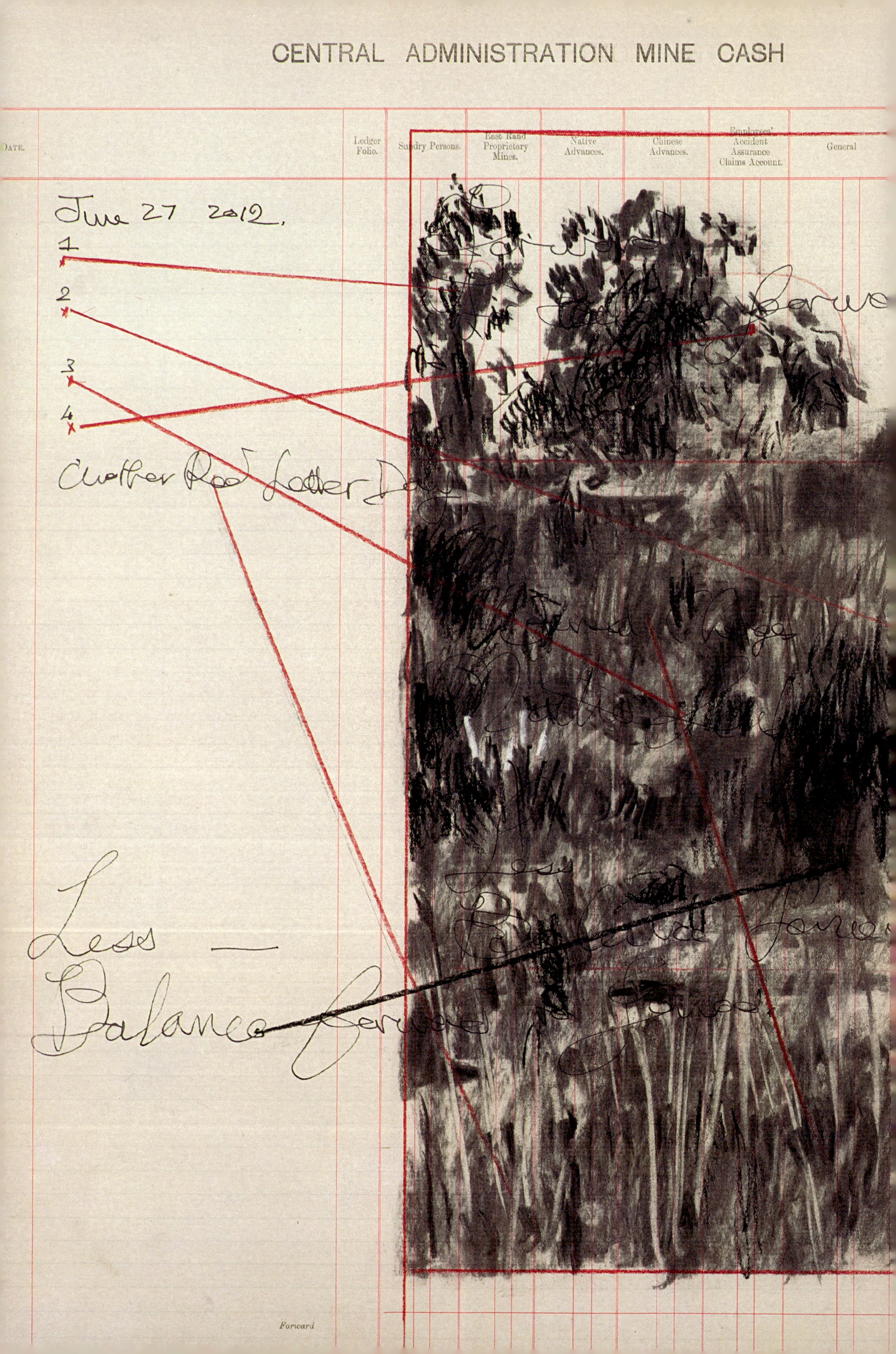
CENTRAL ADMINISTRATION MINE CASH
Date.
Ledger Folio.
Sundry Persons.
East Rand Proprietary Mines.
Native Advances.
Chinese Advances.
Employees' Accident Assurance Claims Account.
General
June 27 2012.
1
2
3
4
Less
Balance Forward
Forward

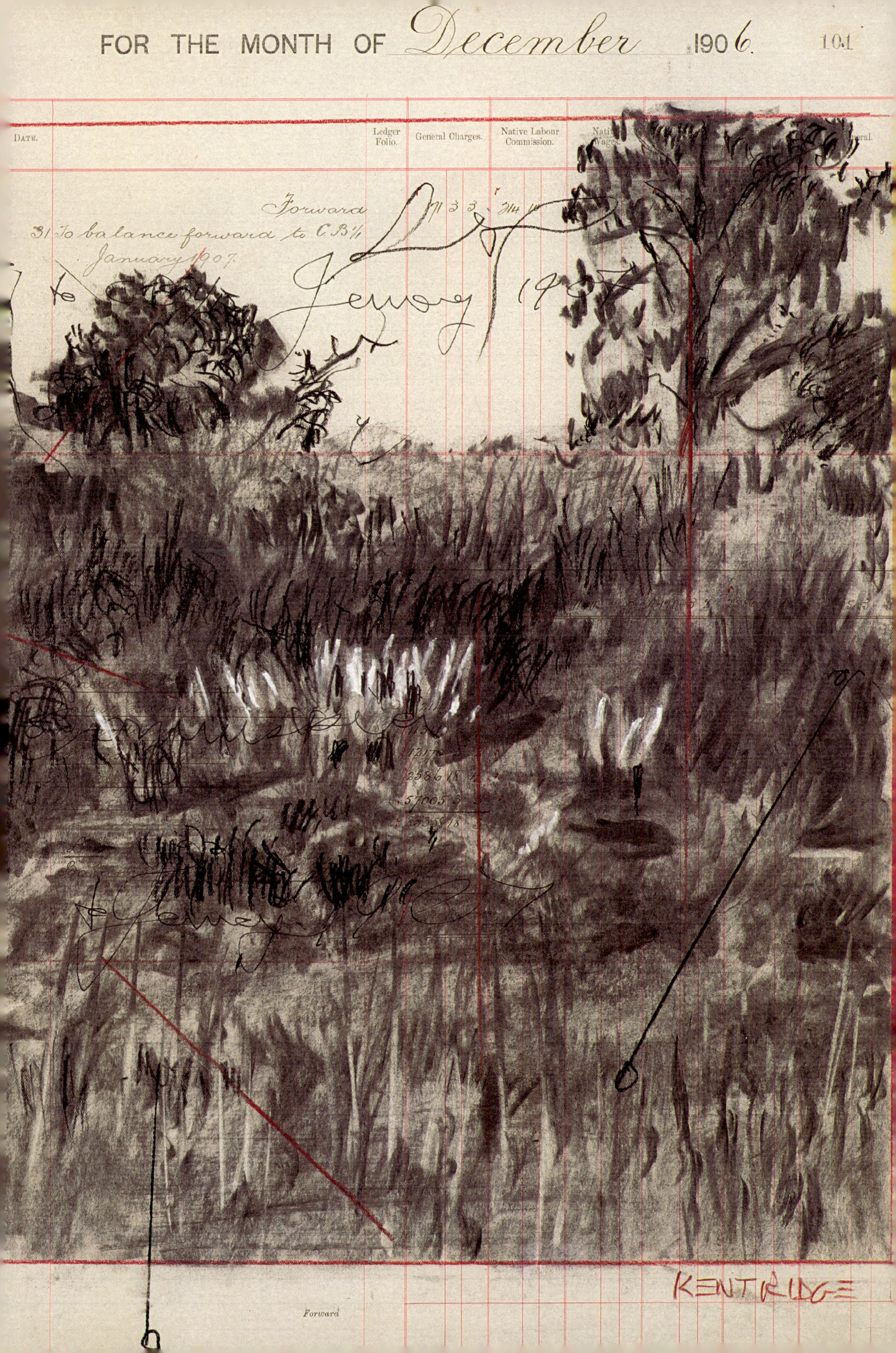
FOR THE MONTH OF December 1906
101
Date.
Ledger Folio.
General Charges.
Native Labour Commission.
Forward
31 To balance forward to C.B.1/1
January 1907
KENTRIDGE
Forward

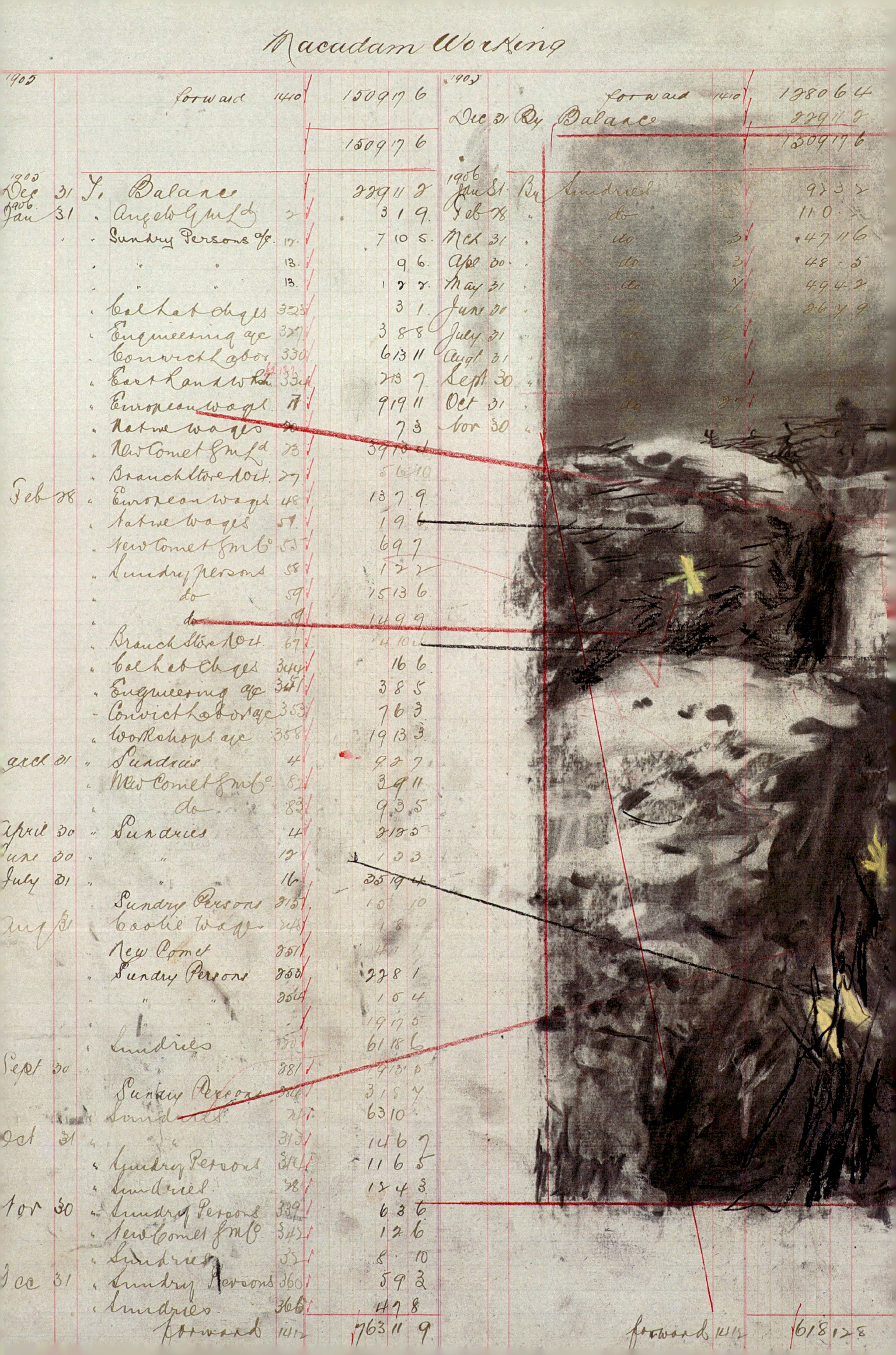

Macadam Working
forward 1410 1509 17 6
Dec 31 By Balance
1906 Jan 31 By Sundries

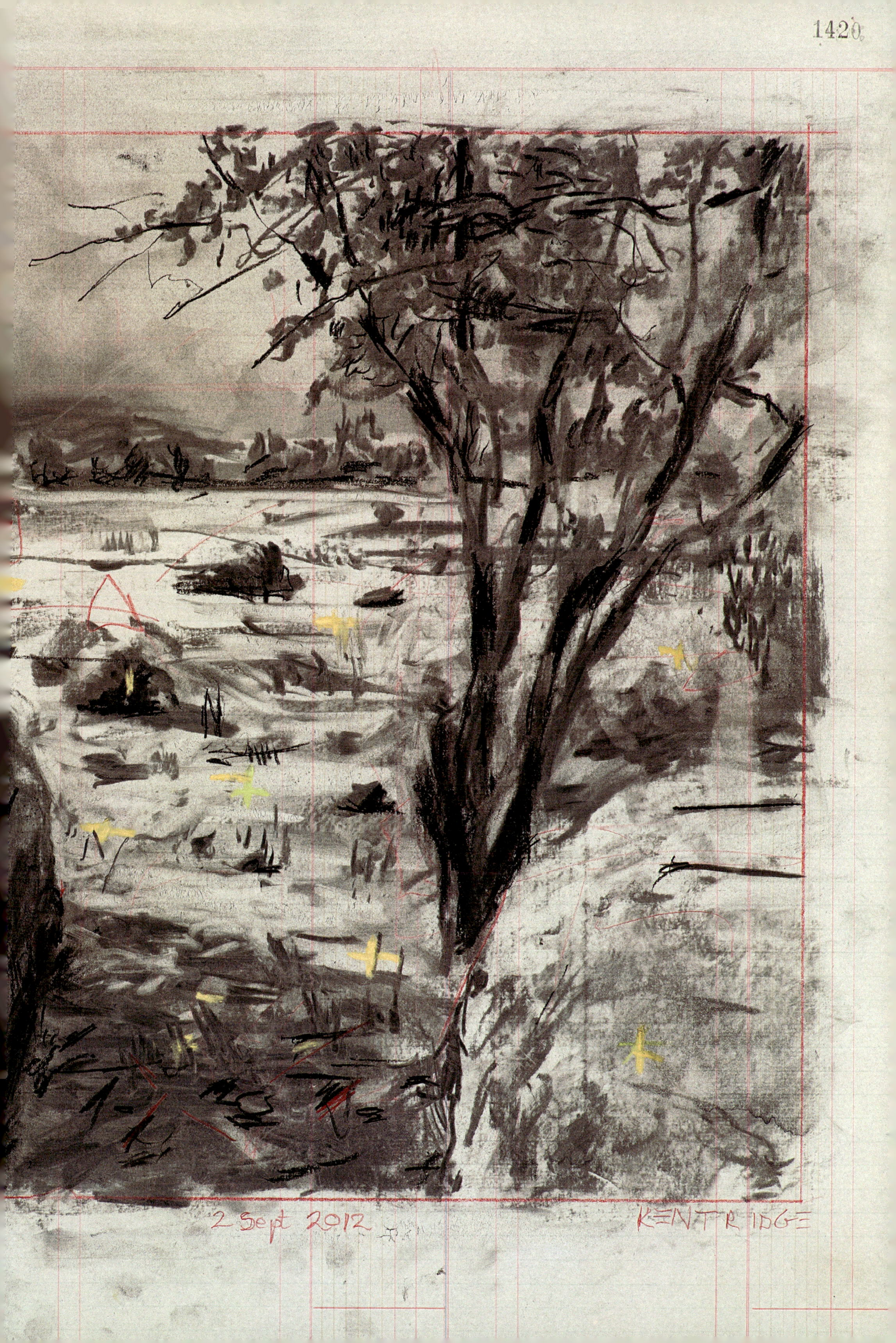
1420
2 Sept 2012
KENTRIDGE

CENTRAL ADMINISTRATION MINE CASH

Date		Ledger Folio.	Sundry Persons.	East Rand Proprietary Mines.	Native Advances.	Chinese Advances.	Employees' Accident Assurance Claims Account.	General
Aug	Forward		613 11 5	47 . .	48 12 6	1574 10 10	1241 12 4	42976 12 4
22	The Rand Mutual Assce Co Ltd	759	50					
	Coolie No 50376. £10 : - : - ✓							
	✓× 54933 5 : - : - ✓							
	✓ 2080 5 : - : - ✓							
	× 39476. 5 : - : - ×							
	✓ 48168 10 : - : - ✓							
	✓ 38435. 5 : - : - ✓							
	Native No 100542 10 : - : - ✓							
	Native at Driefontein (46501 19 5 / 50 " / Add } 41502 14 5)				3			
	" "				12			
	Contractors Location A/c	842	15 12 6					
	Recd from G. O. of Labour rents collected from Driefontein Location as per letters from D/n Compound Manager dated 24-8-06							
	Levy S.	541	12 6					
	Water A/c July.							
	Amount collected by Compound Manager D/ftein on a/c of advance Native 2793 as per letter 9th Aug 1906				9 6			
	Chinese Capitation Fees.	363						6 10
	Chinese Passage Fees	369						5
	Chinese Repatriation Expenses	561						6
	Imported Labour Com. Gen. Exp.	781						2 6
	Recd from Liu Hai Chuan No 2132 who arrived 4 June 1905.							
	Chinese Capitation Fees	363						6 10
	Chinese Passage Fees.	369						5
	Chinese Repatriation Expenses.	561						6
	Imported Labour Com. Gen. Exp.	781						2 6
	Recd from Tuan Shun Ho No 39179 who arrived 4 June 1905.							
	Bolt A.	118	7 6					
	Coal A/c July.							
	Mayberry J.	603	1					
	W & T. A/c Aug							
	Clymo & Co	128	2 12 1					
	March & May A/cs as per Statement.							
	New Transvaal Chemical Co Ltd	629	1 4 6					
	Workshops A/c s/o 20 Dec 1906 Shop Order No 1820 Day Book folio 97.		~~1 4 6~~					
	Innes R. L.	457	14					
	1 Bag Kaffir Corn Main Store A/c Aug.							
	Forward		685 14 6	47 . .	49 17	1574 10 10	1241 12 4	43011 17

Date		Ledger Folio	General Charges	Native Labour Commission	Native Wages	Chinese Wages	Sundry Persons	General
	Forward		7 · 5	536 12 ·	6722 4 ·	9410 13 10	759 7 4	1391 6 8
21	Coolies at Angelo					1327 8 5		
	Natives at Driefontein				2 13 10			
	"				13 10			
	"				12			
	"				3			
	Deceased Native Wages Suspense A/c	688						6 8 2
	Paid Native Commissioner Jo'burg amount due in the estate of deceased natives as per contra. (Cheque 78)							
	Postage Stamps		1					
	Stamps for receipts		1					
	Angelo Gold Mines Ltd	18						15 9
	New Comet G. M. Co. Ltd.	1562						7 9
	Cason Gold Mines Ltd.	390						17 6
	Driefontein Consolidated Mines Ltd	589						15 6
	Prop of G. W. Solomons expenses to town with gold.							
	Natives Return Home A/c	1536						6 13 7
	Cost of 229 natives returning home.							
	Unclaimed European Wages A/c	(1927)						
	Paid for Death Certificate G. Galley	1927						5
	Taylor J. E.	879					84 · 3	
	As per J. V.							
	New Comet G. M. Co. Ltd.	1562						10
	Paid W. K. Thompson as per Journal Voucher							
	McMullin & Still	601					40	
	On a/c of Contract							
	Exchange on £25000		15 12 6					
	Chinese Compound Chgs Angelo	376						74 10
	Paid E. Cain Interpreter passage money from East Rand to China. (Distribute over 3 mos.							
	Employees Accident Assce Claims A/c	759						36 7 3
	Paid the following in full settlement of their claims for injuries sustained							
	James W. H. £13 · 6 · 4							
	Eschdorf W. 8 : 10 : -							
	Jory W. 7 : - : -							
	Knox S. 4 : 1 : 4							
	Stevens E. J. £2 ; Davies J. £1 · 9 · 7 ; 3 : 9 : 7							
	Forward		24 12 11	536 12 ·	6726 6 8	10738 2 3	883 7 7	1598 7 2

CENTRAL ADMINISTRATION MINE CASH

Date.		Ledger Folio.	Sundry Persons.	East Rand Proprietary Mines.	Native Advances.	Chinese Advances.	Employees' Accident Assurance Claims Account.	General
	Forward							

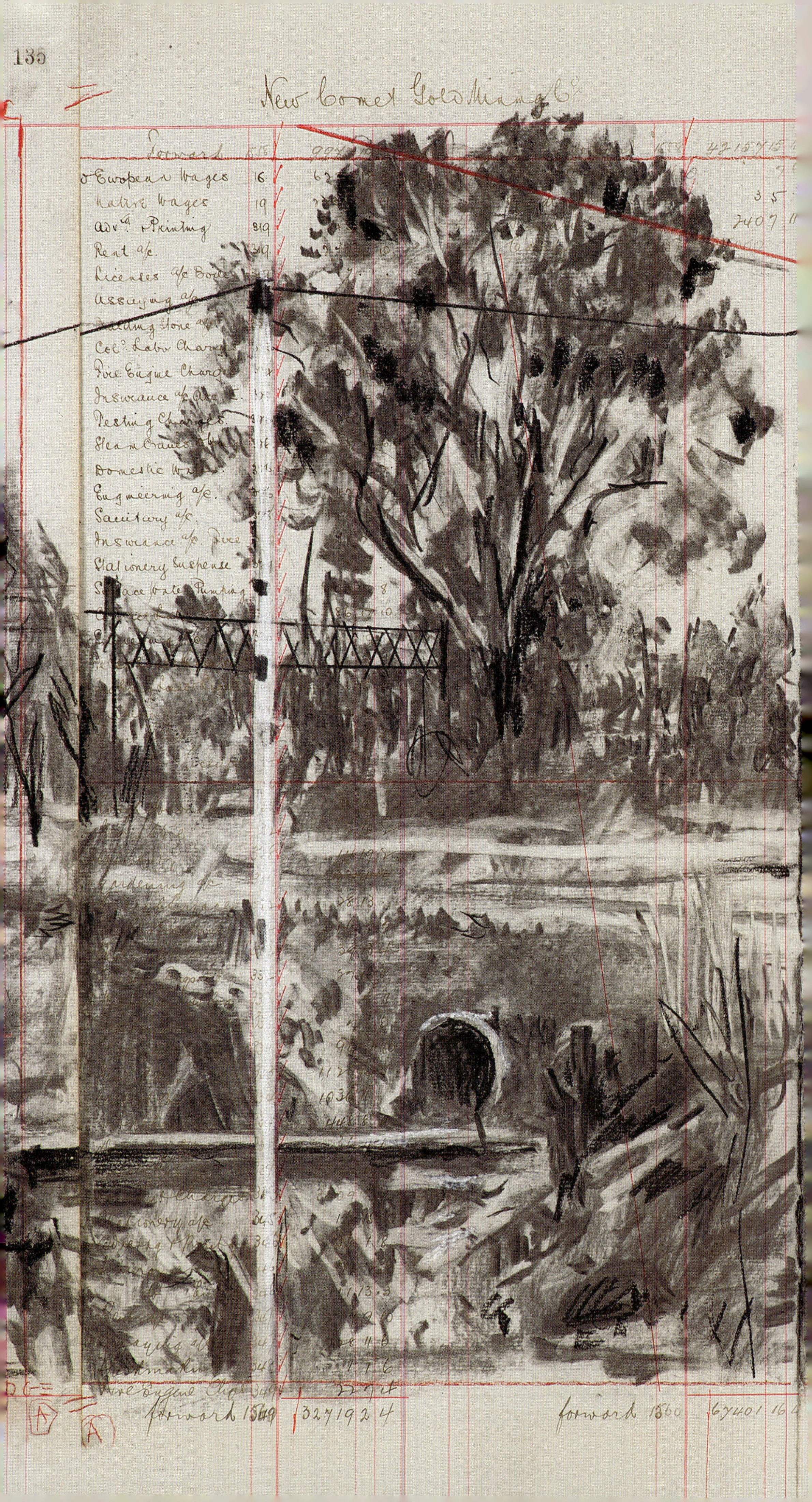

135
New Comet Gold Mining Co
Forward
European Wages
Native Wages
Advt. Printing
Rent a/c
Assaying a/c
Col. Labr Charges
Fire Engine Charges
Testing Charges
Domestic
Engineering a/c
Sanitary a/c
Stationery Suspense
forward 1549
3271924
forward 1560
67401

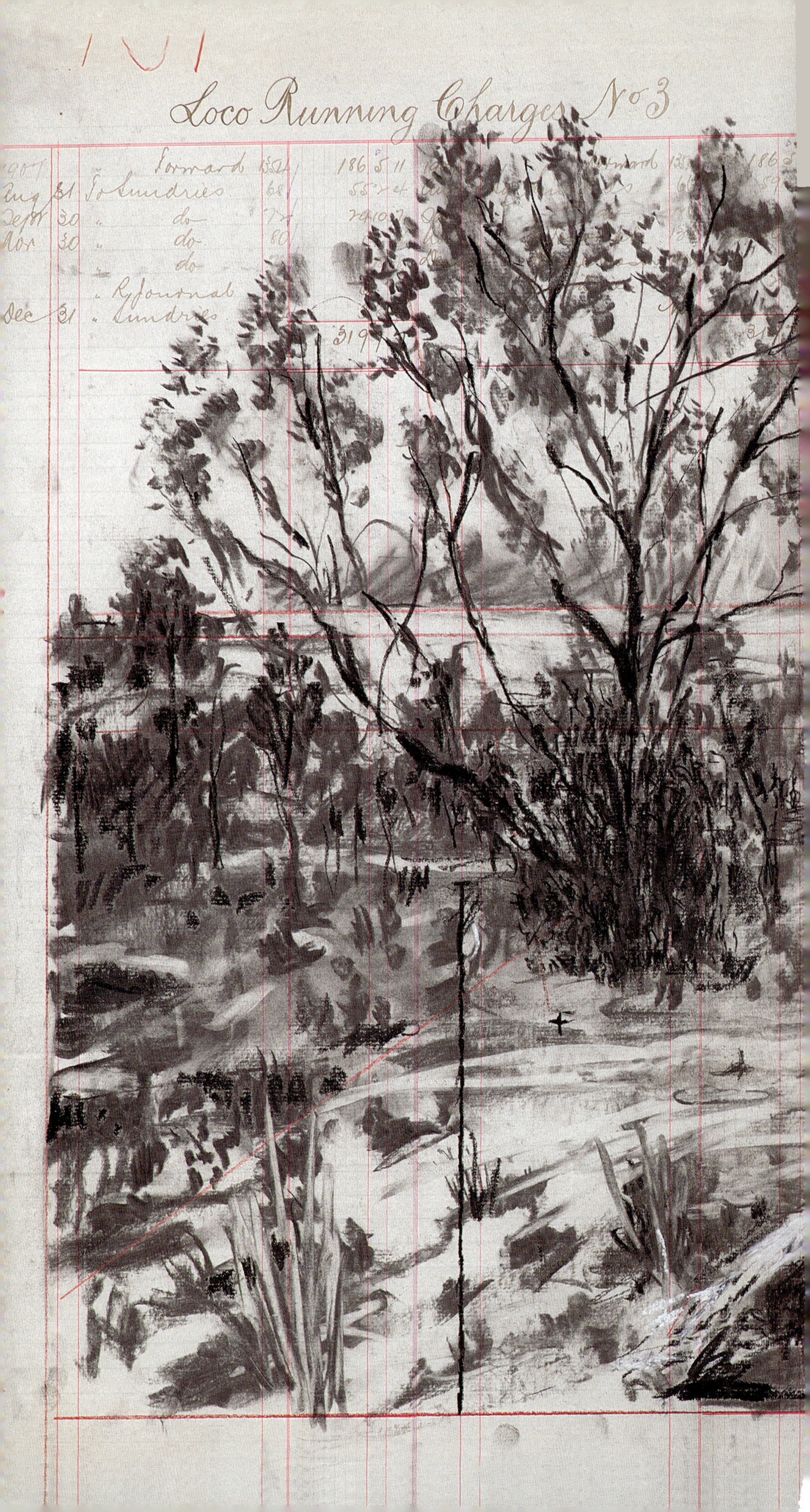
101
Loco Running Charges No 3
Forward
Aug 31 To Sundries
Dept 30 do
Nov 30 do
do
Journal
Dec 31 Sundries

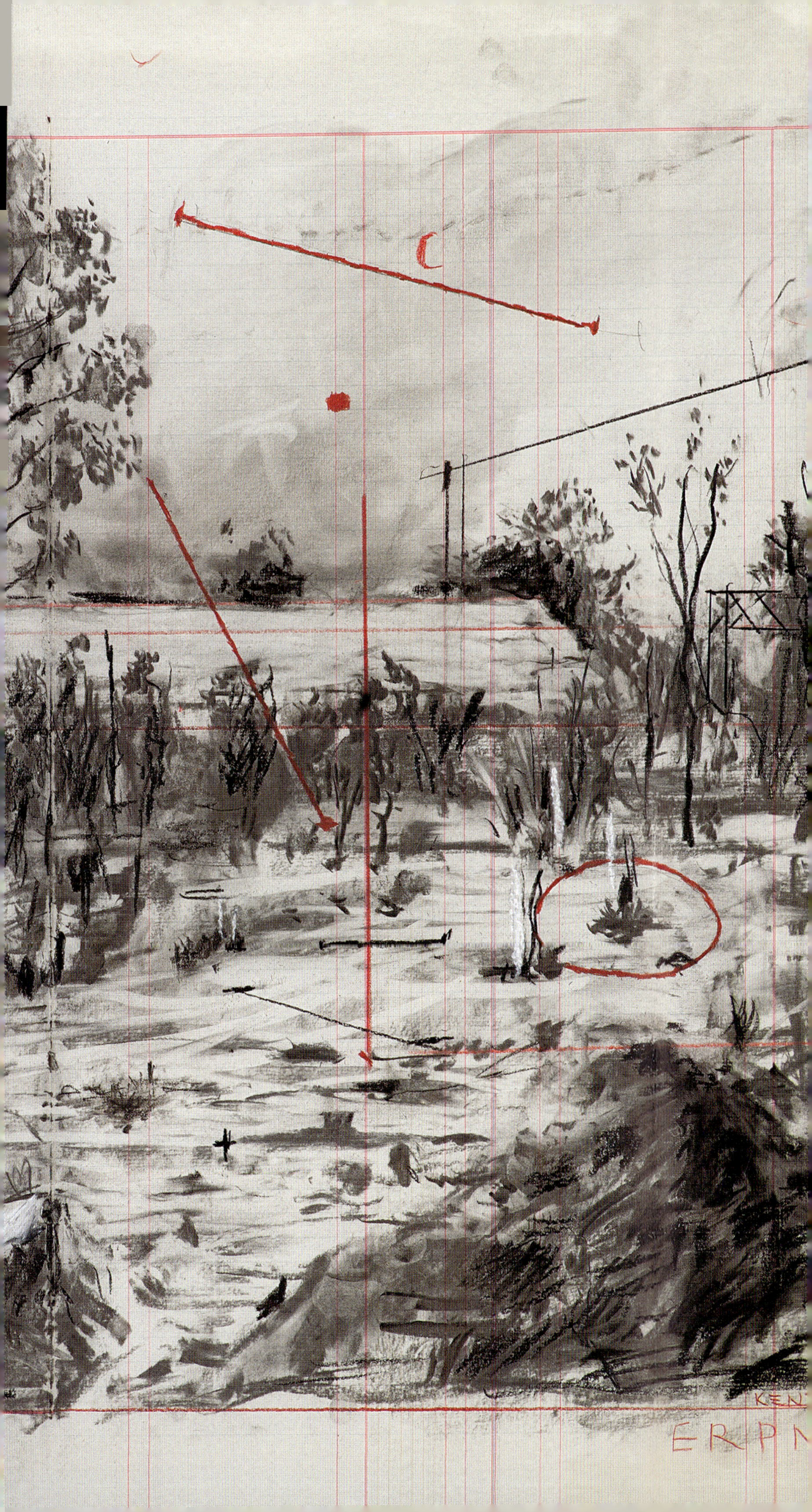

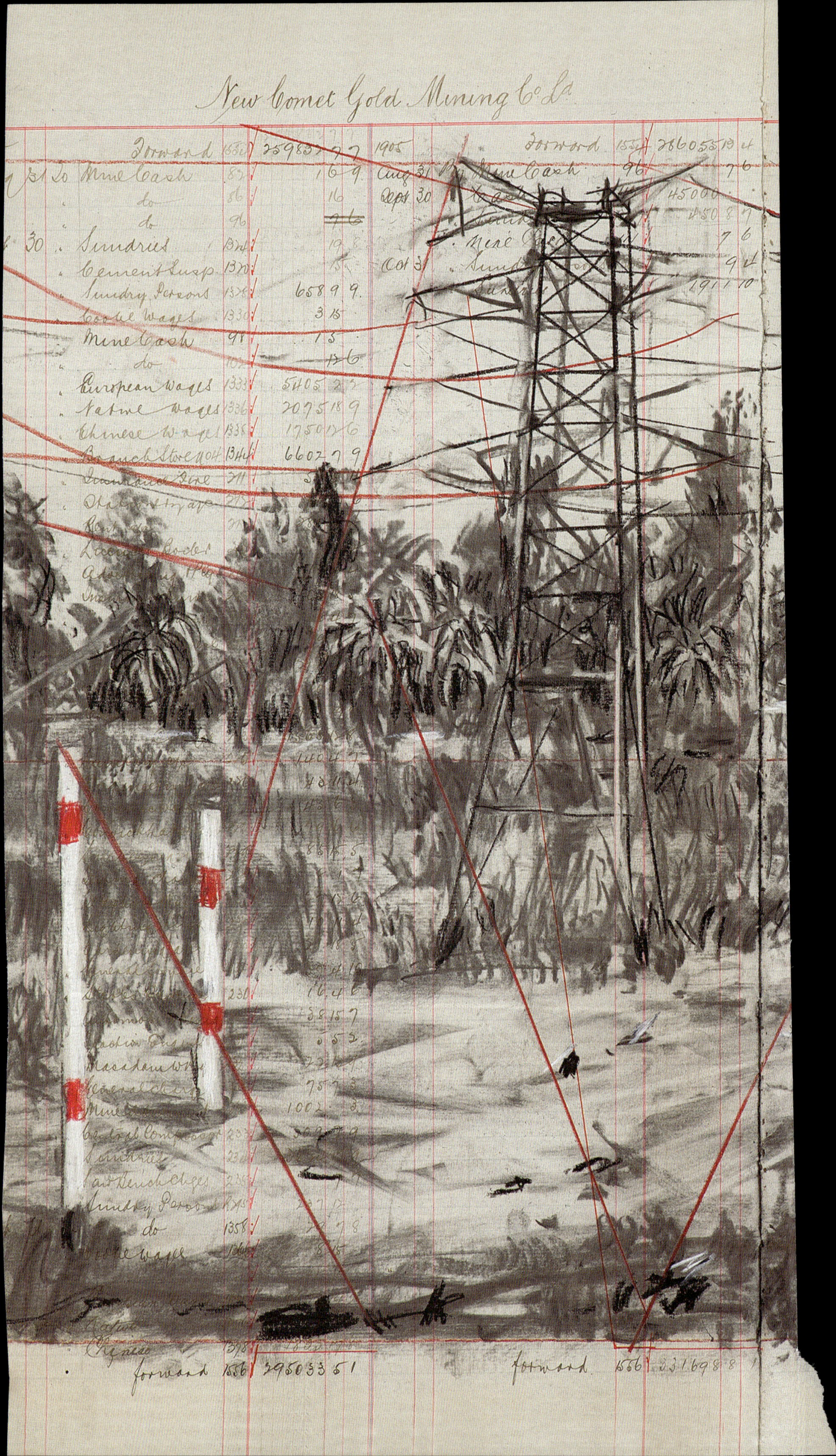
New Comet Gold Mining Co Ld
Forward
Mine Cash
Sundries
Cement Susp
Sundry Persons
Coolie Wages
European Wages
Native Wages
Chinese Wages
Branch Store
forward
forward

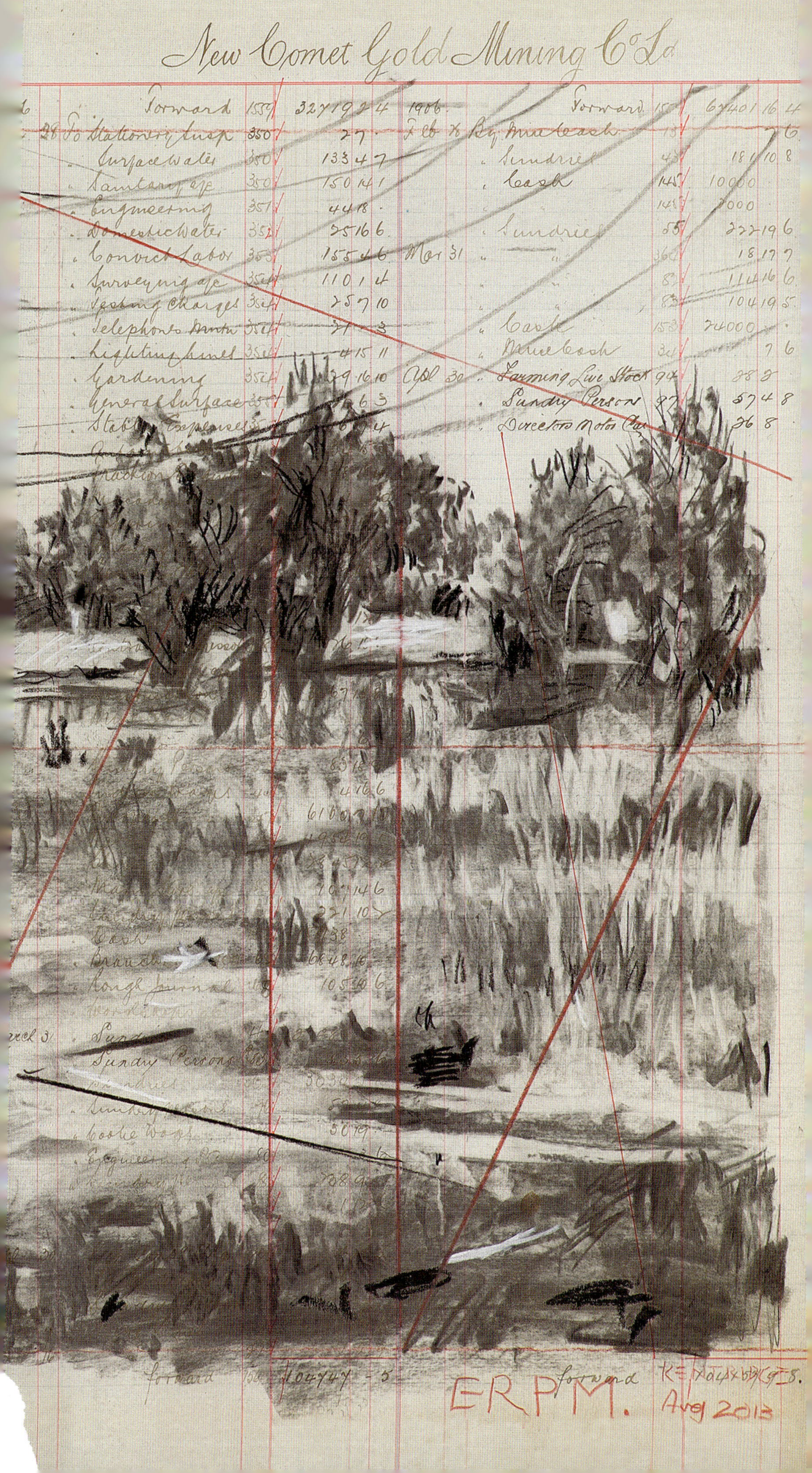
New Comet Gold Mining Co Ld
ERPM.
Aug 2013

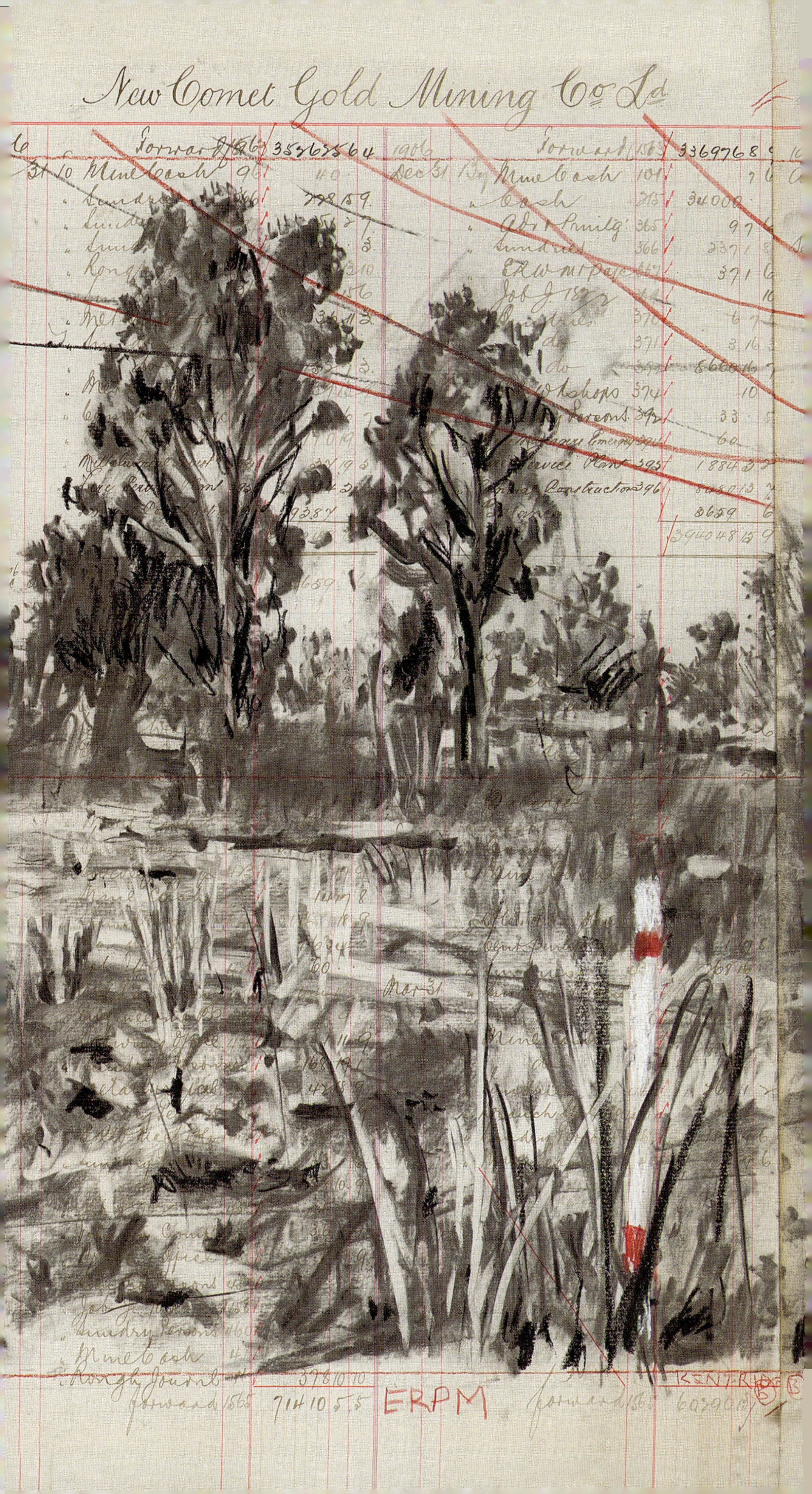
New Comet Gold Mining Co Ld
ERPM

FOR THE MONTH OF 190

Date.		Ledger Folio.	General Charges.	Native Labour Commission.	Native Wages.	Chinese Wages.	Sundry Persons.	General.
	Forward							

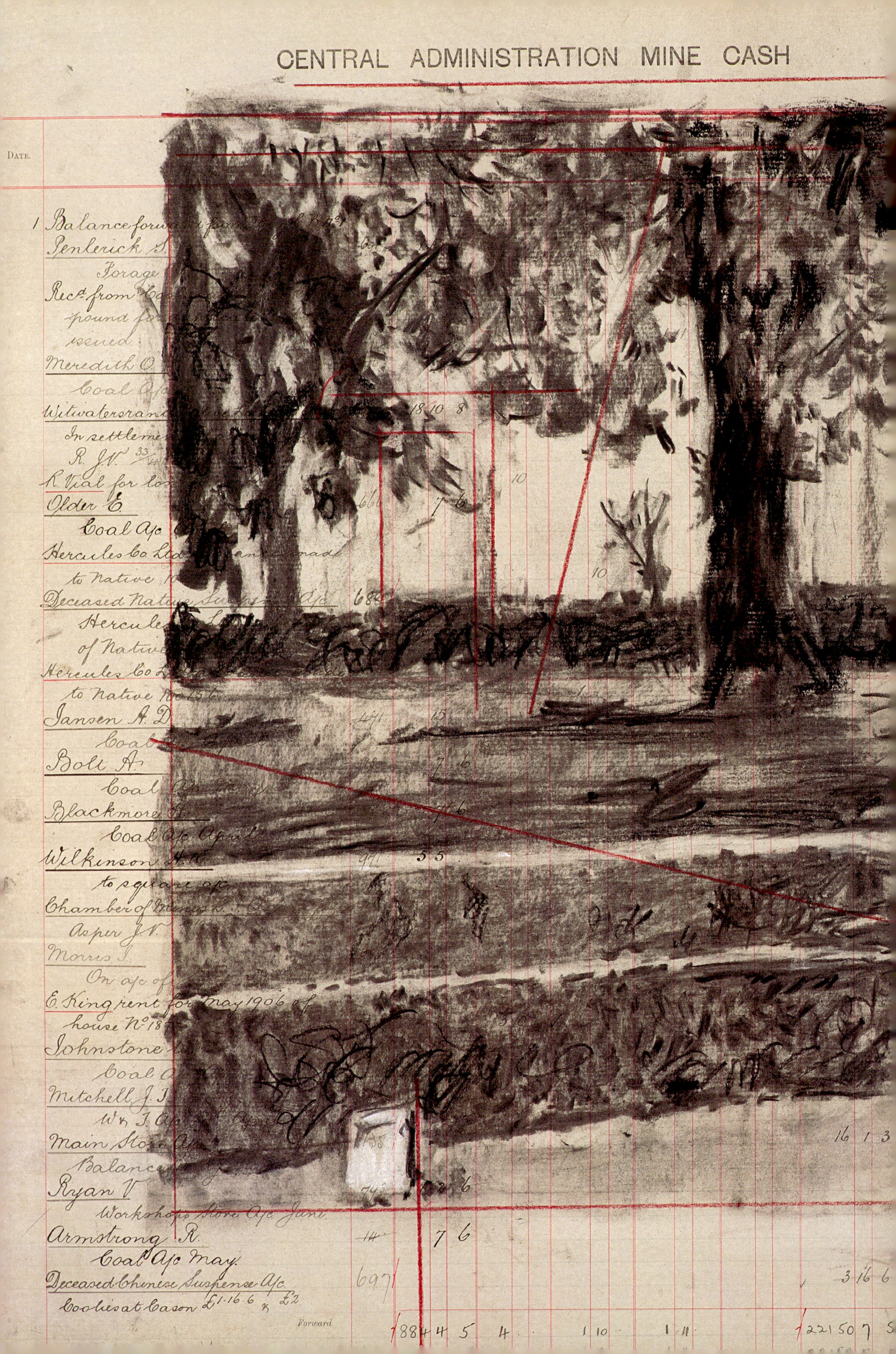
CENTRAL ADMINISTRATION MINE CASH
DATE.
Older E.
Coal A/c
Jansen A. D
Bolt A.
Armstrong R.
Coal A/c May
Deceased Chinese Suspense A/c
Forward

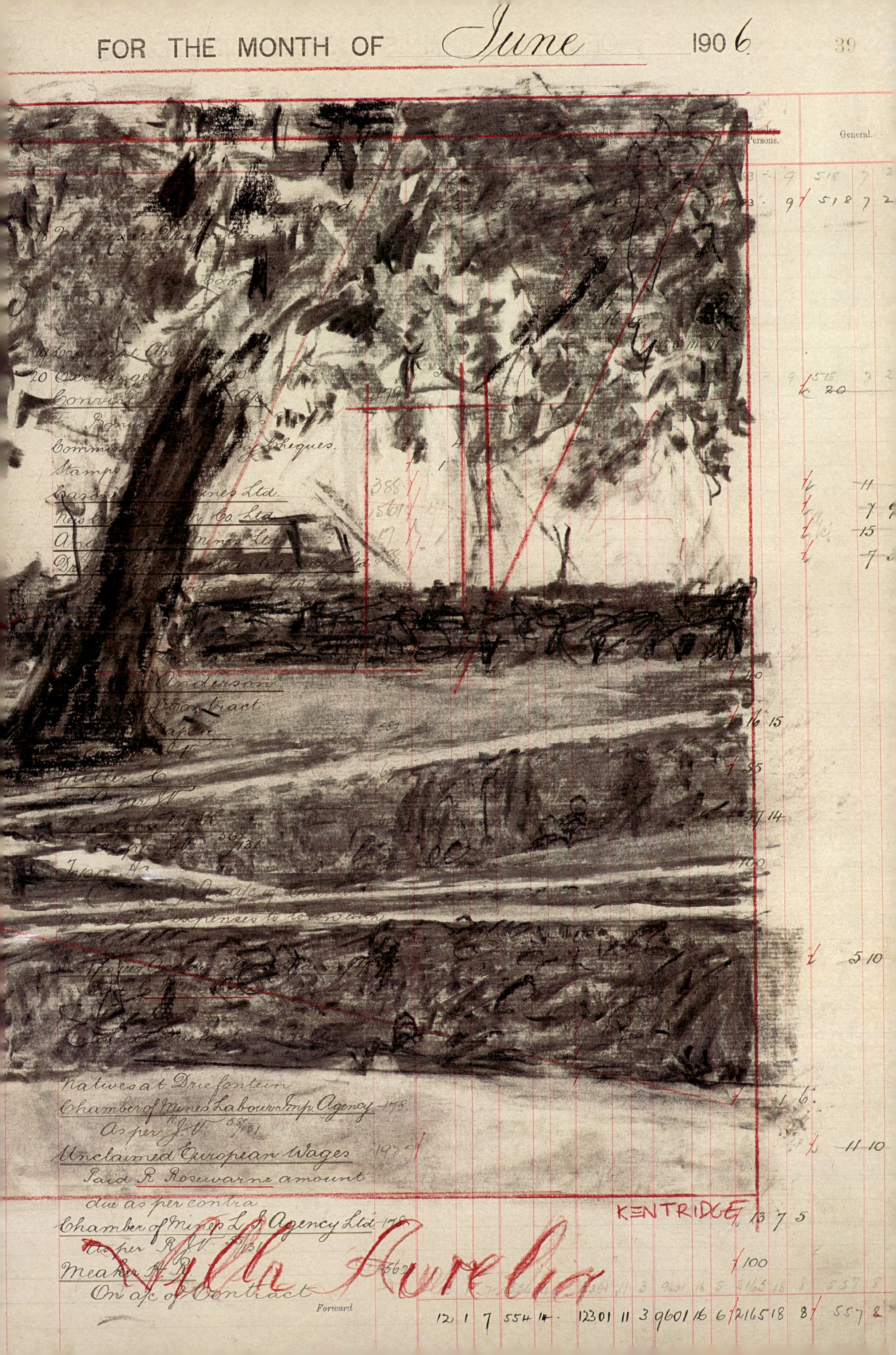
FOR THE MONTH OF June 1906
39
General.
Natives at Driefontein
Chamber of Mines Labour Imp. Agency
As per J.V.
Unclaimed European Wages
Paid R. Rosewarne amount
due as per contra
Chamber of Mines L. I. Agency Ltd
KENTRIDGE
Villa Aurelia
On a/c of Contract
Forward

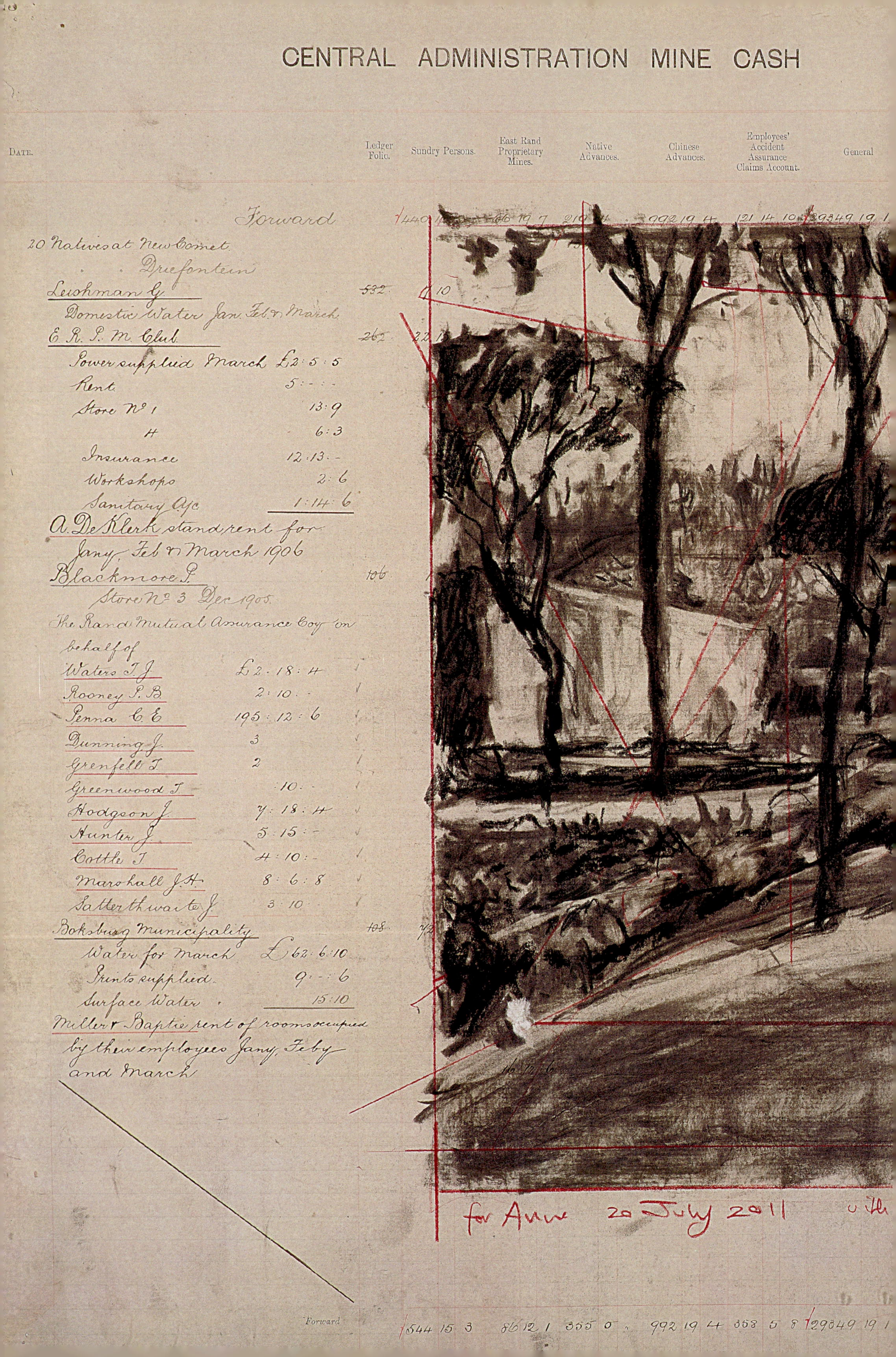

CENTRAL ADMINISTRATION MINE CASH
DATE.
Ledger Folio.
Sundry Persons.
East Rand Proprietary Mines.
Native Advances.
Chinese Advances.
Employees' Accident Assurance Claims Account.
General
Forward
20 Natives at New Comet Driefontein
Leishman G.
Domestic Water Jan. Feb. & March
E. R. P. M. Club
Power supplied March £2:5:5
Rent 5:-:-
Store No 1 13:9
4 6:3
Insurance 12:13:-
Workshops 2:6
Sanitary A/c 1:14:6
A. De Klerk stand rent for Jany Feb & March 1906
Blackmore S.
Store No 3 Dec 1905
The Rand Mutual Assurance Coy on behalf of
Waters T. J. £2:18:4
Rooney P. B. 2:10
Penna C. E. 195:12:6
Dunning J. 3
Grenfell T. 2
Greenwood T. 10
Hodgson J. 7:18:4
Hunter J. 5:15:-
Cottle T. 4:10:-
Marshall J. H. 8:6:8
Satterthwaite J. 3:10
Boksburg Municipality
Water for March £62:6:10
Prints supplied 9:-:6
Surface Water 15:10
Miller & Baptie rent of rooms occupied by their employees Jany, Feby and March
for Anne 20 July 2011
Forward
544 15 3
86 12 1
355 0
992 19 4
368 5 8
129849 19 1

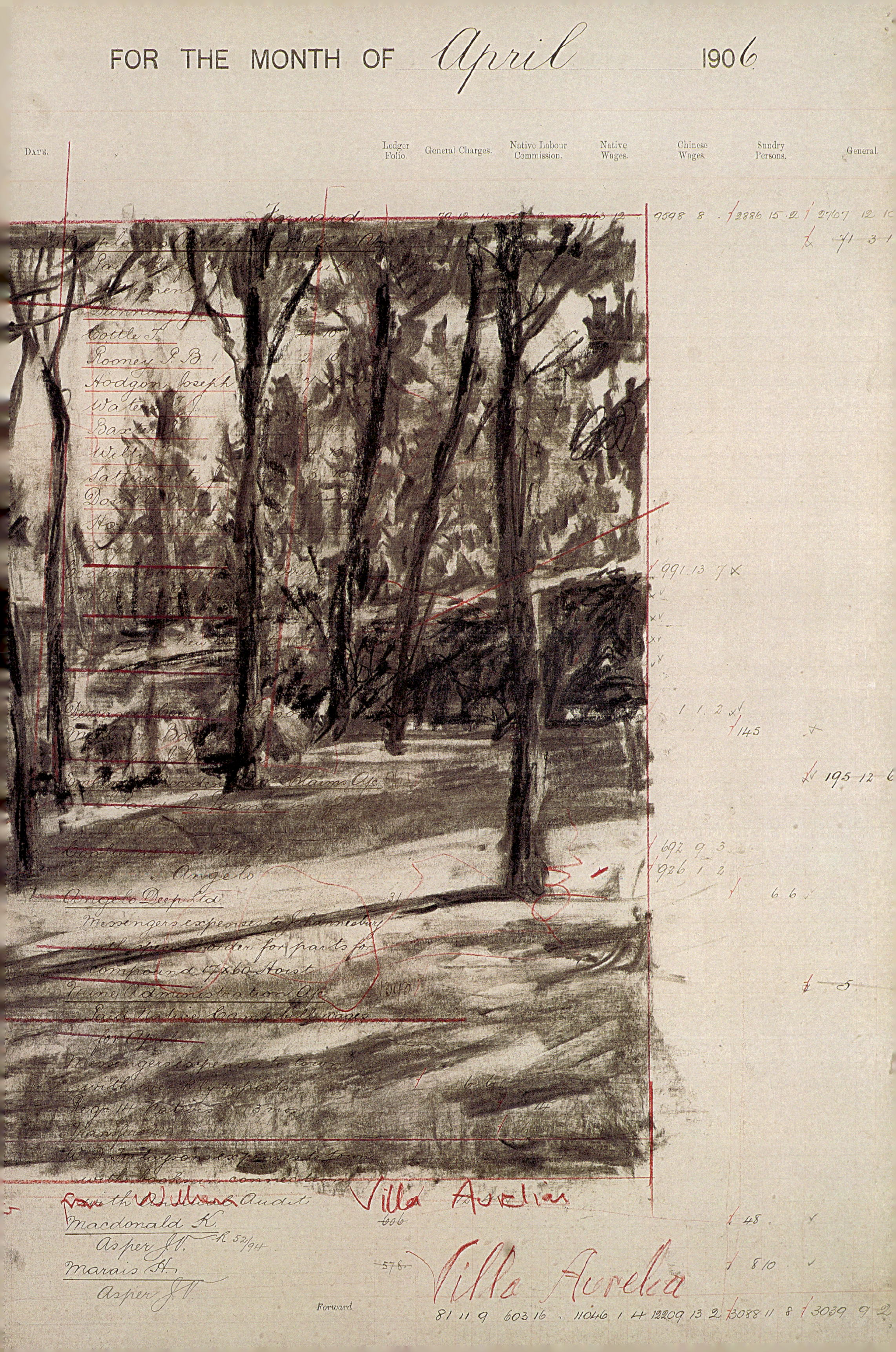

FOR THE MONTH OF April 1906
Date.
Ledger Folio.
General Charges.
Native Labour Commission.
Native Wages.
Chinese Wages.
Sundry Persons.
General.
Forward
Villa Aurelia

CENTRAL ADMINISTRATION MINE CASH

Date.		Ledger Folio.	Sundry Persons.	East Rand Proprietary Mines.	Native Advances.	Chinese Advances.	Employees' Accident Assurance Claims Account.	General.
	Forward		900 11 2	5 15 –	36 15 6	75 19 3		22281 8 3
12	Hook G.D.	431	1 – 6					
	Main Store A/c May £ : 13 –							
	Coal A/c : 7 6							
	Coolies at (Angelo) Comet 3rd Arrivals					– 9 6		
	" " " " Clothing					3 – –		
	Chinese Defaulters A/c	570						– 2 –
	Coolies at (Angelo) Comet							
	Coolies at Angelo					22 15 11		
	" " " (Clothing)					2 14 –		
	Chinese Defaulters A/c	573						– 7 3
	Coolies at Angelo							
	Chinese Remittance A/c	533						92 15 –
	Coolies at Angelo							
	Coolies at Angelo repatriated					2 7 –		
	" " " "					– 6 9		
	Chinese Defaulters A/c	573						~~8 X X~~ 19 9
	Coolies at Angelo £ : ~~7 X~~ 7 : 11							
	" " " £ : 11 : 10							
	Chinese Remittance A/c	533						2 10 3
	Coolies at Angelo £ 1 : 18 : 1							
	12 : 2							
	Coolies at Angelo repatriated					2 17 9		
	Deceased Native Wages Suspense A/c	688						11 14 6
	Native at Driefontein £ 1 : 1 : –							
	" " " 3 : – : 6							
	" " " 2 : – : 6							
	" " " 1 : 12 : –							
	" " " 2 : 8 : 6							
	1 : 12 : –							
	Lishman G.	532	2 10 –					
	Water A/c May							
	Wilson W.H.M.	961	– – 5					
	to balance a/c							
13	Recd from G. Blair Hook, G.O. of Labour				2 4 –			
	" " " " " " " "				1 11 –			
	" " " " " " " "				2 8 –			
	Contractors Location A/c	842	~~13 – 6~~ 8					
	Recd from G.B. Hook							
	Recd from G.B. Hook					5 ~~6~~ 4 3		
	Natives Return Home A/c	1537						2 12 6
	Recd from G.B. Hook							
	Native Labour Com.	1530						1 17 2
	Recd from G.B. Hook for Natives purchasing their Discharge as per letter dated 13 June 1906.							
	Forward		912 2 1	5 15 –	42 18 6	116 – 5		22394 6 8

FOR THE MONTH OF June 1906

Date.		Ledger Folio.	General Charges.	Native Labour Commission.	Native Wages.	Chinese Wages.	Sundry Persons.	General.
			3 19 10	554 14	847 14 3	4557 12 1	1620 14 3	447 16 2
	Forward		3 19 10	554 14 ·	847 14 3	4557 12 1	1620 14 3	447 16 2
12	J. Langs expenses to town re Von Schade & Nortons case		8 6					
	Unclaimed Native Wages.	1931						10
	Paid J Lang on a/c of Native Thomas Mine No 958 Reg No 15852.							
	Messengers expenses to town with Weekly Reports		6 6					
	Meaker H. R.	562					100	
	On a/c of Contract							
	Miller C.	614					60	
	On a/c of Contract							
	Angelo Deep Ltd.	31					6 6	
	Messengers expenses to town with special letter to Secretary re Cabling Development Results.							
	Convict Labour A/c	479						19
	Bonuses to Convict Warders at following Places:-							
	Driefontein Battery £2 : - : -							
	" Crusher 3 : - : -							
	Angelo " 2 : - : -							
	Comet Battery 2 : - : -							
	" Crusher 2 : - : -							
	Phoenix Sorting 1 : - : -							
	Cason Crusher 1 : - : -							
	" Surface Tramming 1 : 10 : -							
	" " " 1 : - : -							
	Coal Depot 1 : 10 : -							
	Driefontein Coal 1 : - : -							
	Cason Slimes Dams 1 : - : -							
	Deceased Coolie at New Comet					1 3		
	8213 : 3 : 3		4 14 10	554 14	847 14 3	4557 13 3	1781 · 9	467 6
	Natives at Driefontein Hill.				524 17 4			
	" " New Comet				1833 11 6			
	" " "				234 5 ·			
	" " " (B. C. A.)				18 19 6			
	" " "				3 7 4			
	" " "				4 11 6			
	" " "				3 16 8			
	" " "				2208 18 9			
	Coolies at Cason.					24 17 5		
	" " " 7th Arrivals					1707 11 3		
			4 14 10	554 14	5680 1 10	6290 1 11	1781 · 9	467 6
	Forward		4 14 10	554 14	5680 1 10	6290 2 ·	1781 · 9	467 6 ·

CENTRAL ADMINISTRATION MINE CASH

Date		Ledger Folio.	Sundry Persons.	East Rand Proprietary Mines.	Native Advances.	Chinese Advances.	Employees' Accident Assurance Claims Account.	General.
	Forward		1058 7 -	1865 17 1	430 7 9	1315 19 4	763 2 10	97549 18 6
30	Boarding House Licences A/c	1034						3 18 9
	T. J. Hill £ : 12 : 6							
	M. Thompson : 12 : 6							
	J. T. Davies : 12 : 6							
	M. Faull (3 mos) 2 : 1 : 3							
	Boarding House Com. A/c	707						33 5 -
	J. T. Davies ½% on £1879 : 18 : 6 £9 : 7 : 6							
	T. J. Hill ½% on £1687 : 14 : 11 8 : 7 : 6							
	M. Thompson ½% on £1084 : 1 : 11 5 : 7 : 6							
	M. Faull ½% on £250 : 17 : - 1 : 10 : -							
	M. Morrison 1% on £863 : 5 : 1 8 : 12 : 6							
			1058 7 -	1865 17 1	430 7 9	1315 19 4	763 2 10	97587 2 3
	Sundry Persons	2015	1058 7 -					
	East Rand Prop Mines Ltd	778	1865 17 1					
	Native Advances	1542	430 7 9					
	Chinese Advances	475	1315 19 4					
	Employees Accident Assce Claims A/c	761	763 2 10					
	General	✓	97587 2 3					
			103020 16 3					
	Less							
	Balance forward from March £4608 : 5 : 3							
	Head Office £92000 : - : -		96608 5 3					
		1447	6412 11 -					

FOR THE MONTH OF April 1906

Date.		Ledger Folio.	General Charges.	Native Labour Commission.	Native Wages.	Chinese Wages.	Sundry Persons.	General.
	Forward		1 . 9	570 17 6	2190 19 5	2627 4 11	1735 4 3	278 8 1
11	Coolies at Cason					635 1 11		
	" "					883 3 7		
	Natives at D & A.				175 6 3			
	" " Driefontein				996 4 10			
	Coolies at New Comet					705 1 3		
	" " Angelo					15 6		
	" " " (Interpreters)					50		
						1012 13 8		
	" " "					~~2023 3 5~~		
12	Natives at New Comet				2007 14 2			
	Meaker H. R.	562					120	
	On a/c of Contract							
	Convict Labour A/c	479						46
	Bonuses for month of March							
	Chief Warder £5 : - : -							
	2 Overseers 5 : - : -							
	4 " 8 : - : -							
	3 Warders 6 : - : -							
	4 " 6 : - : -							
	Phoenix Shaft Sorting 1 : 10 : -							
	Comet Crusher Station 2 : - : -							
	" Mill 2 : - : -							
	Angelo Crusher & Sorting 2 : - : -							
	Driefontein " " 3 : - : -							
	" Mill 2 : - : -							
	E. R. P. M. Coal 1 : 10 : -							
	Driefontein " 1 : - : -							
	Cason Slimes Dam 1 : - : -							
	Macdonald. Kenneth	606					50	
	As per J. Voucher R 67/81							
	Boksburg Municipal Licences	1303						65
	4 Traction Engines (3 mos) £20 : - : -							
	30 " Trailers (3 mos) 45 : - : -							
	Boiler Licences	1335						3 6 8
	Licences for Boilers 7948, 7949 7950, 7951 (10 mos)							
	Godwin & Anderson	371					200	
	On a/c of Contract							
	Unclaimed Chinese Wages	1979						5 2 3
	Paid Coolies at Angelo (amount included in £31-9-6 C.B. No 1 fol 11)							
	Natives Return Home A/c	1537						– 19 3
	33 Natives fares @ 7d							
	Native Compound Chgs. New Comet	1677						2 10
	Wages to Kaffir Woman for brewing Kaffir Beer							
	Expenses sending boys to W.N.L.A. & P. Curator			7				
	Forward		1 . 9	571 4 6	5370 4 8	5914 . 10	2105 4 3	401 6 3

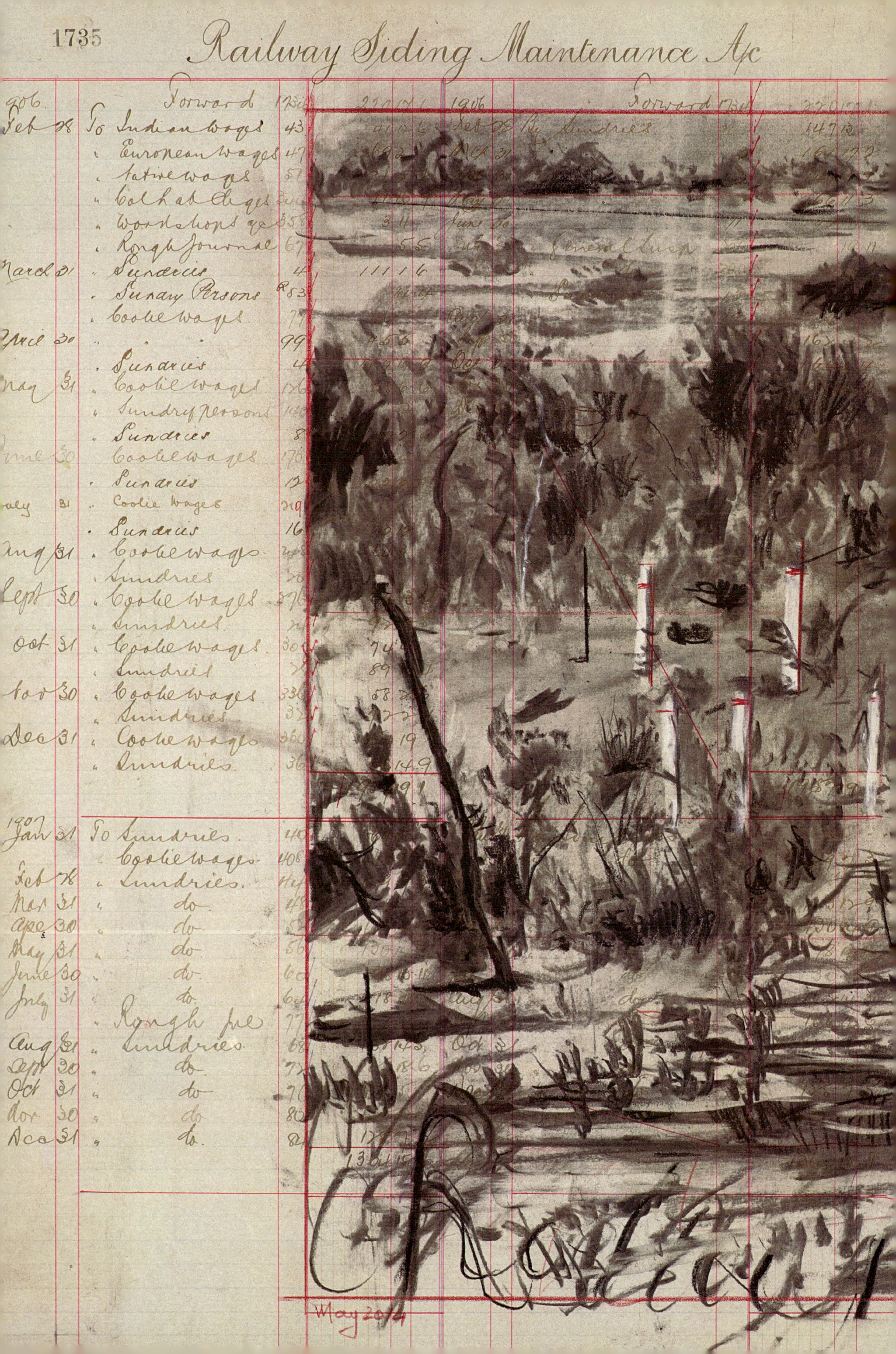
1735
Railway Siding Maintenance A/c
May 2012

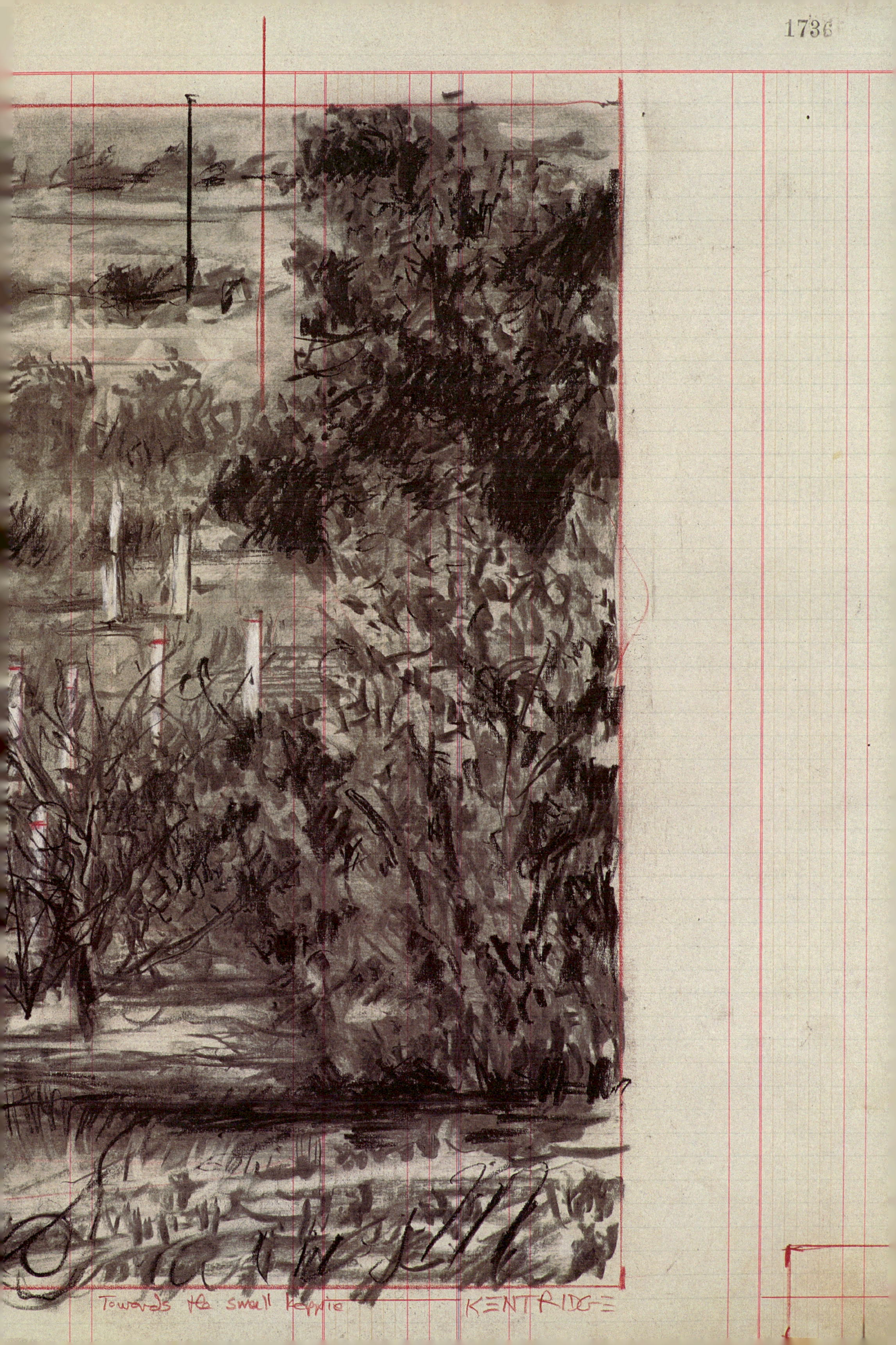
1736
Towards the small koppie
KENTRIDGE

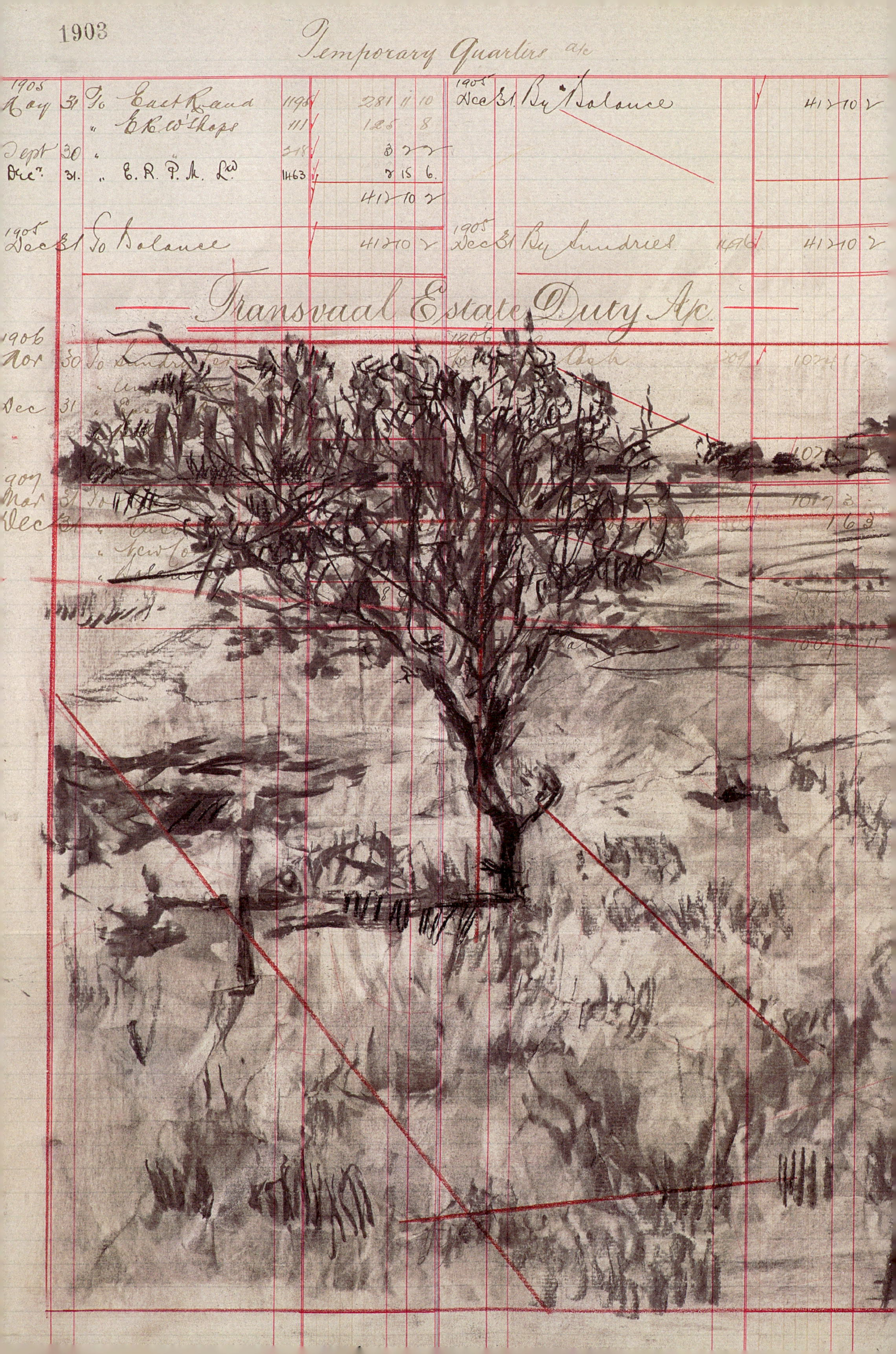
1903
Temporary Quarters a/c
Transvaal Estate Duty A/c

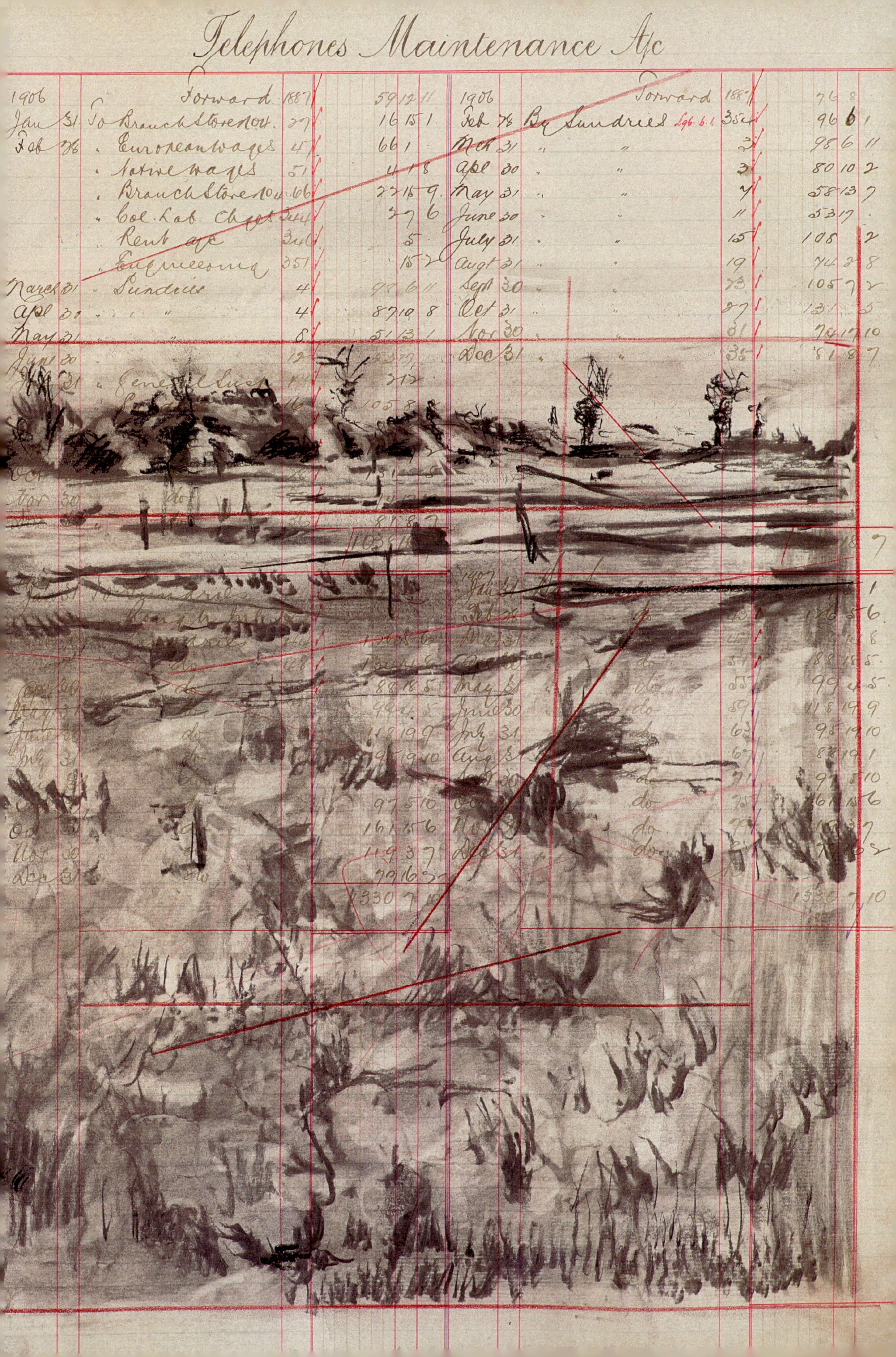
Telephones Maintenance A/c
1906 Forward 1887 59 12 11
Jan 31 To Branch Store No 4 27 16 15 1
Feb 28 European wages 47 66 1
Native wages 51 4 1 8
1906 Forward 1887 76 8
Feb 28 By Sundries 35 96 6 1
Mch 31 3 95 6 11
Apl 30 3 80 10 2
May 31 7 58 13 7
June 30 11 53 17
July 31 15 108 2
Augt 31 19 74 3 8
Sept 30 23 105 7 2
Oct 31 27
Nov 30 31
Dec 31 35 81 8 7

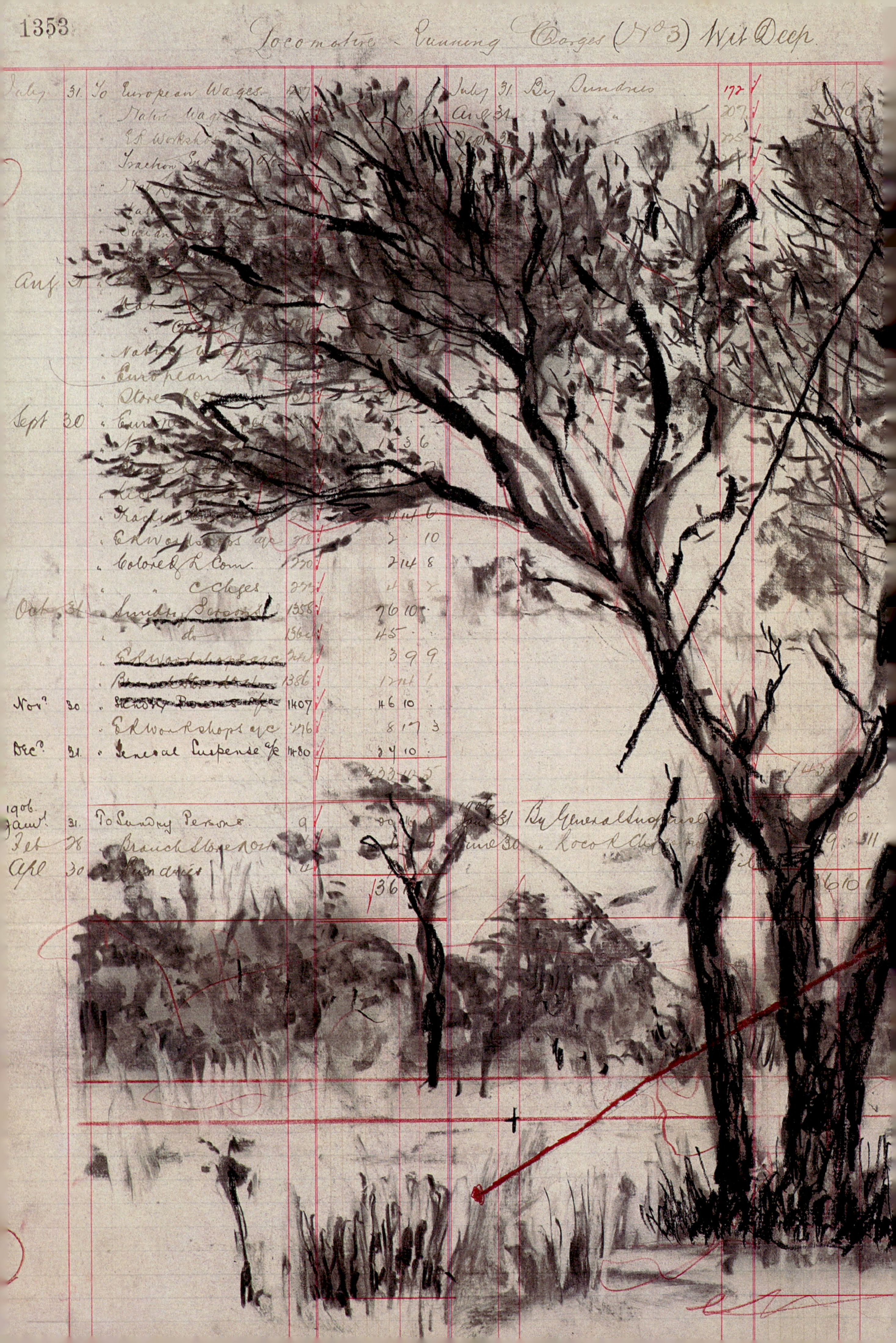
1353
Locomotive Running Charges (No 3) Wit Deep

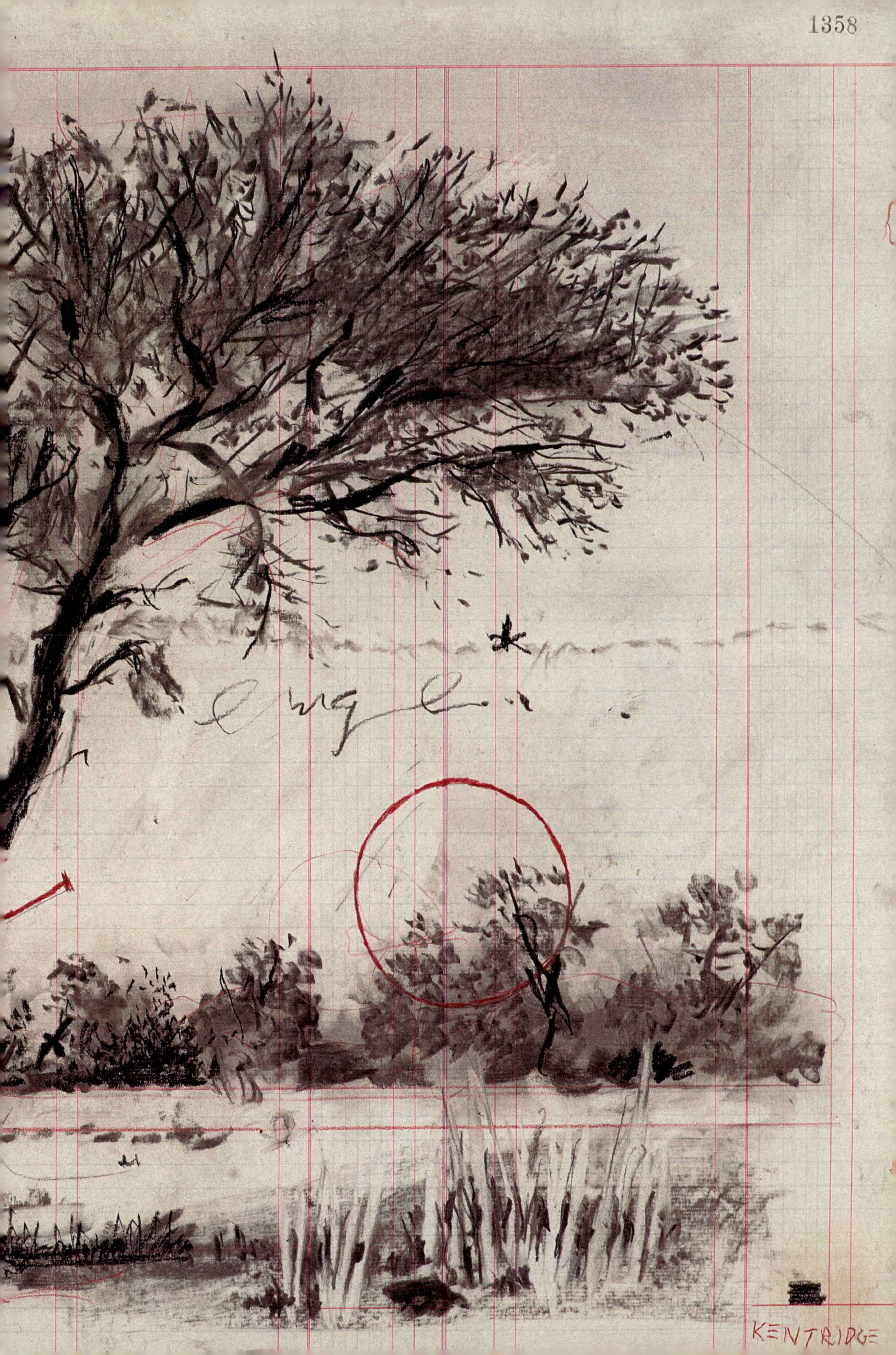
1358
KENTRIDGE

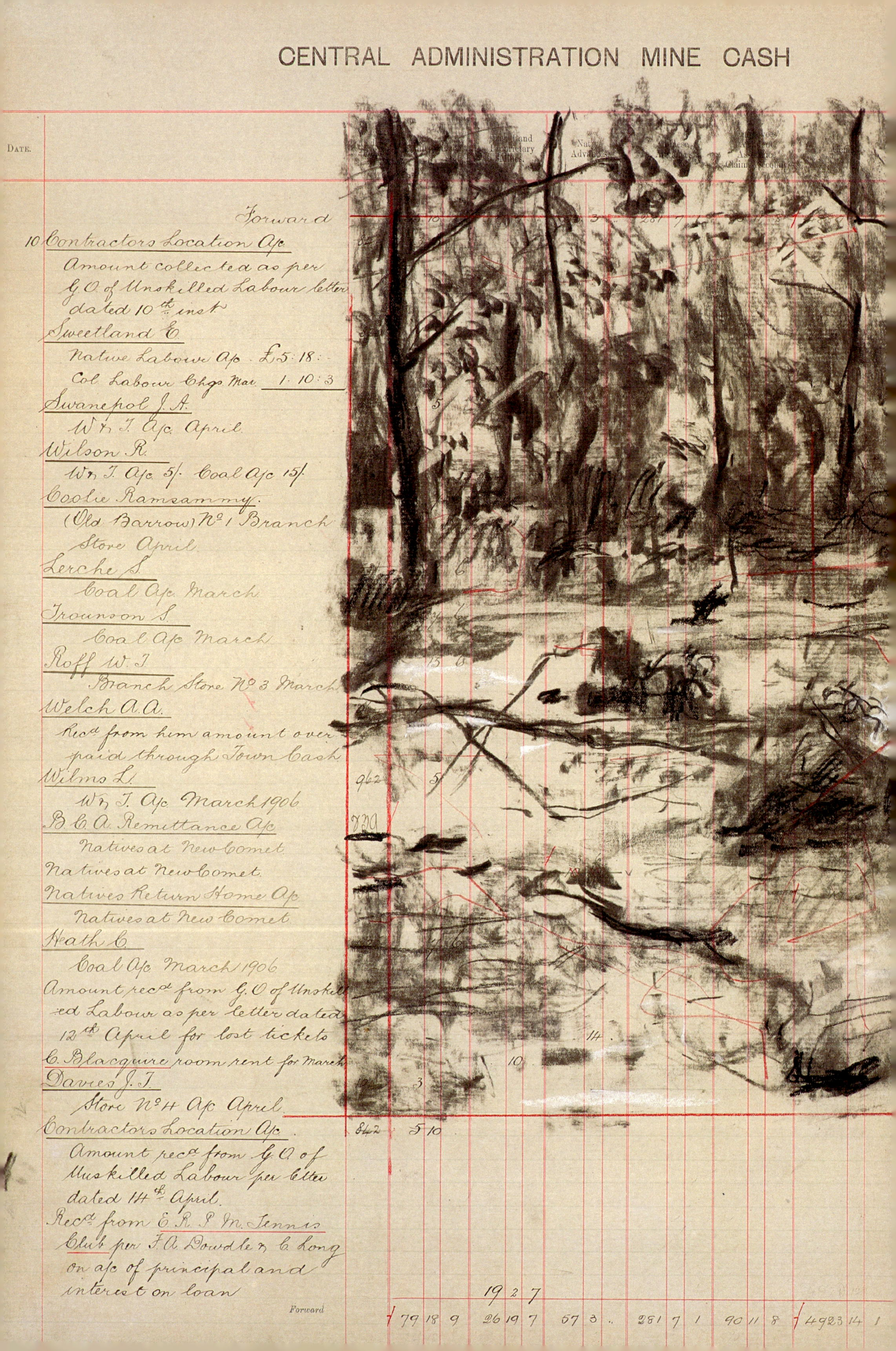
CENTRAL ADMINISTRATION MINE CASH
DATE.
Forward
10 Contractors Location A/c
Amount collected as per
G.O. of Unskilled Labour letter
dated 10th inst
Sweetland E.
Native Labour A/c £5:18:
Col. Labour Chgs Mar. 1:10:3
Swanepol J.A.
W. & T. A/c April
Wilson R.
W. & T. A/c 5/- Coal A/c 15/-
Coolie Ramsammy
(Old Barrow) No 1 Branch
Store April
Lerche S.
Coal A/c March
Trounson S.
Coal A/c March
Roff W.T.
Branch Store No 3 March
Welch A.A.
Recd from him amount over
paid through Town Cash
Wilms L.
W. & T. A/c March 1906
B.C.A. Remittance A/c
Natives at New Comet
Natives at New Comet
Natives Return Home A/c
Natives at New Comet
Heath C.
Coal A/c March 1906
Amount recd from G.O. of Unskill
ed Labour as per letter dated
12th April for lost tickets
C. Blacquire room rent for March
Davies J.T.
Store No 4 A/c April
Contractors Location A/c
Amount recd from G.O. of
Unskilled Labour per letter
dated 14th April
Recd from E.R.P.M. Tennis
Club per F.A. Dowdle & C. Long
on a/c of principal and
interest on loan
Forward
1927
79 18 9 | 26 19 7 | 57 3 .. | 281 7 1 | 90 11 8 | 4923 14 1

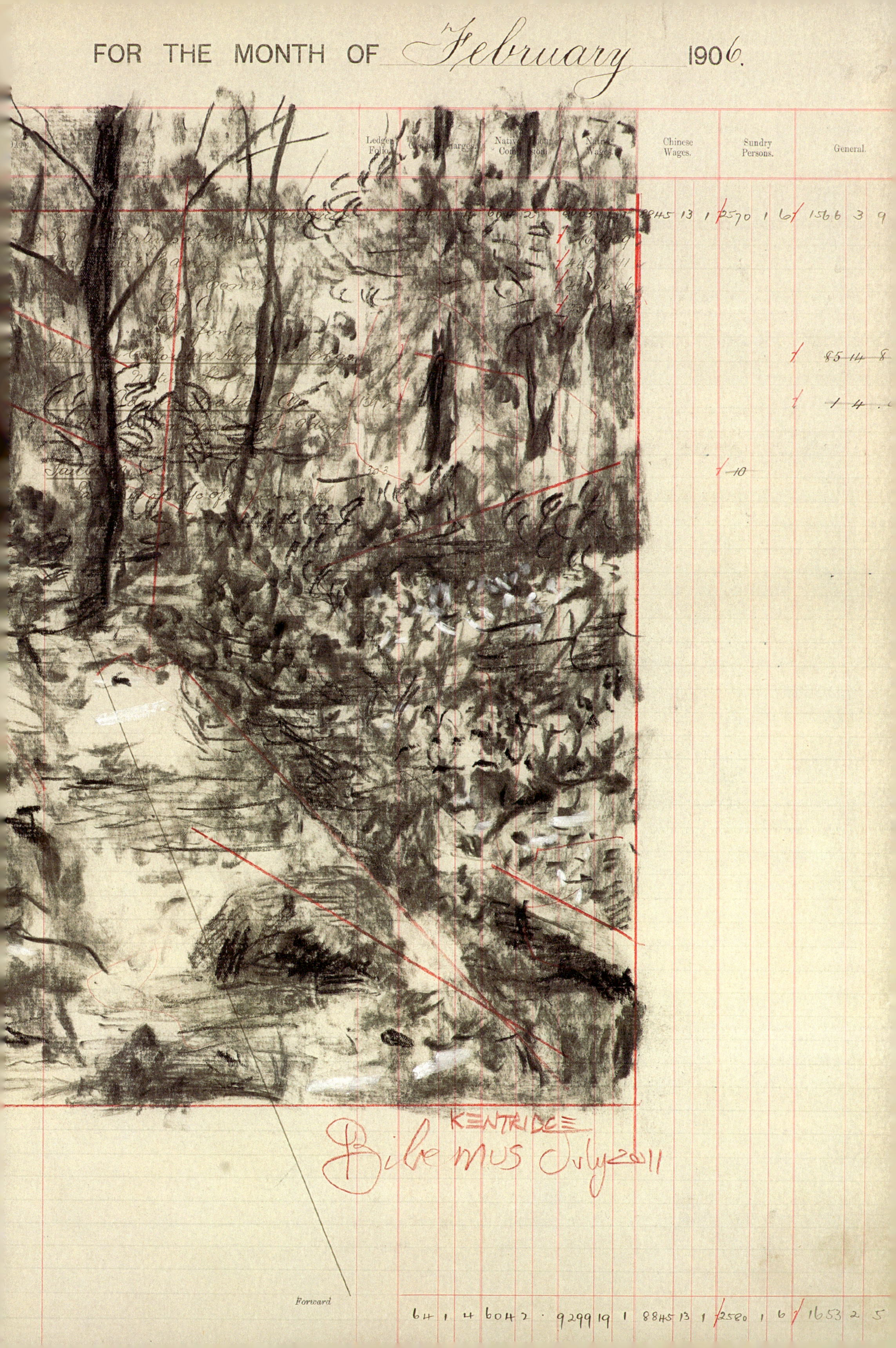
FOR THE MONTH OF February 1906.
Chinese Wages.
Sundry Persons.
General.
Forward
KENTRIDGE
Bibemus July 2011

CENTRAL ADMINISTRATION MINE CASH

Date.		Ledger Folio.	Sundry Persons.	East Rand Proprietary Mines.	Native Advances.	Chinese Advances.	Employees' Accident Assurance Claims Account.	General.
	Forward		437	66 19 7	107 7	551 14 5	121 14 10	13095 10 7
20	Natives Return Home A/c	1537						2 12 6
	Recd from Hercules Co Ltd as per letter dated 20th April							
	Drinkwater F. T.	224						
	Forage A/c Feb & March (Gr. Mealies)							
	Deceased Chinese Suspense A/c	688						3 18 10
	Coolie at Angelo £ : 9 : 10							
	" " " : 4 :							
	" " " April 1 : 5 : 9							
	" " " March 1 : 19 : 3							
	Natives at Driefontein Hill							
	Coolies at Cason							
	" " " (Boots)							
	Chinese Defaulters A/c	578						
	Coolies at Cason							
	Chinese Remittance A/c	533						
	Coolies at Cason							
	Natives Return Home A/c	1537						
	Natives at Driefontein							
	Natives at Driefontein							
	Natives Return Home A/c	1537						
	Natives at New Comet							
	Natives at New Comet							
	B.C.A. Remittance A/c	739						
	Natives at New Comet							
	Coolies at Angelo							
	Chinese Defaulter's A/c	578						
	Coolies at Angelo							
	Chinese Remittance A/c	533						
	Coolies at Angelo							
	Coolies at Angelo							
	Chinese Defaulters A/c	578						
	Coolies at Angelo							
	Coolies at Angelo & Cason 7th Arrivals							
	" " " " " (boots)							
	Chinese Remittance A/c	533						
	Coolies 7th Arrivals							
	Head Office							
	Head Office							
	Forward		440 16 3	66 19 7	219 11 ..	992 19 4	121 14 10	29349 19 1

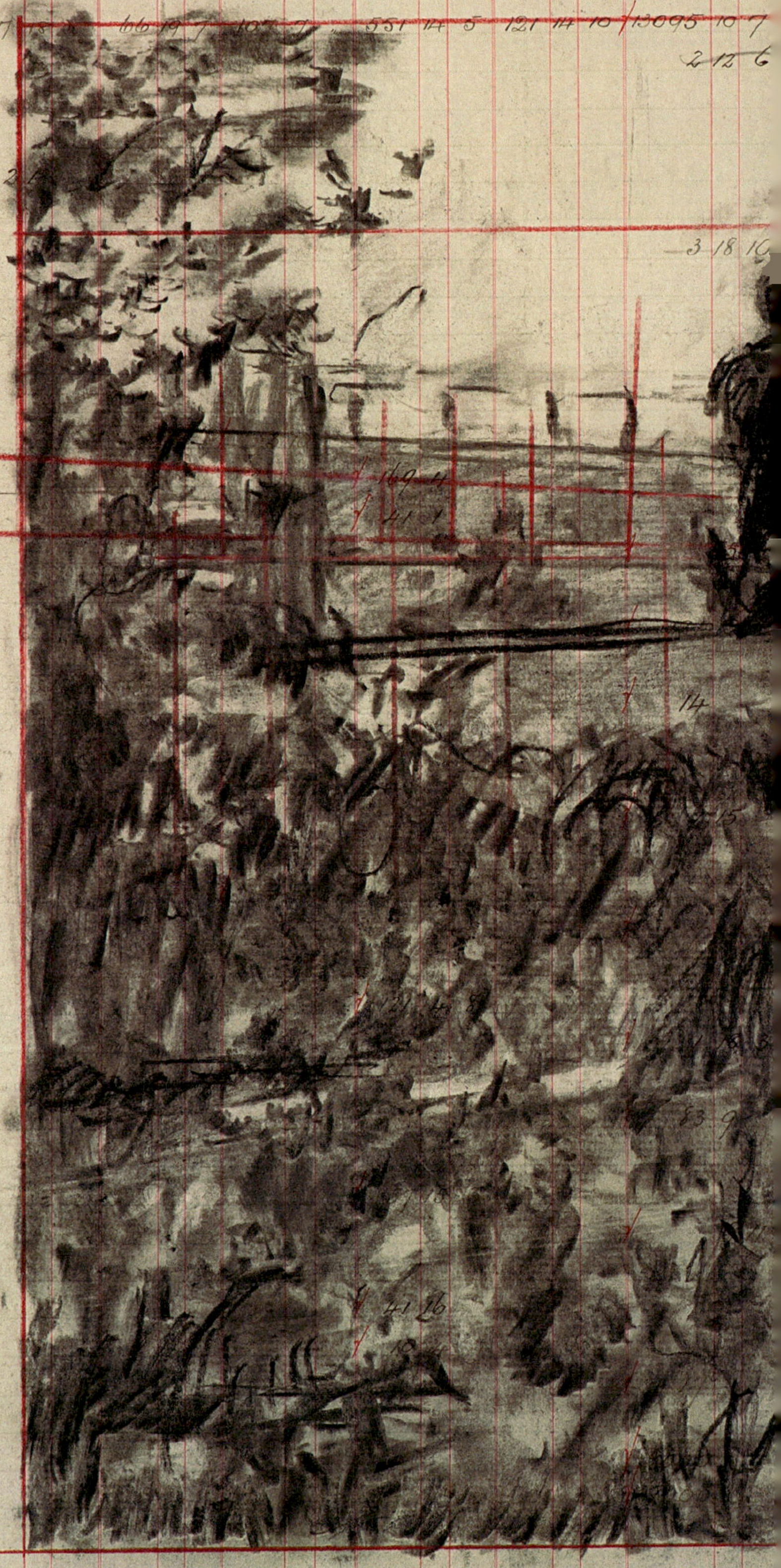

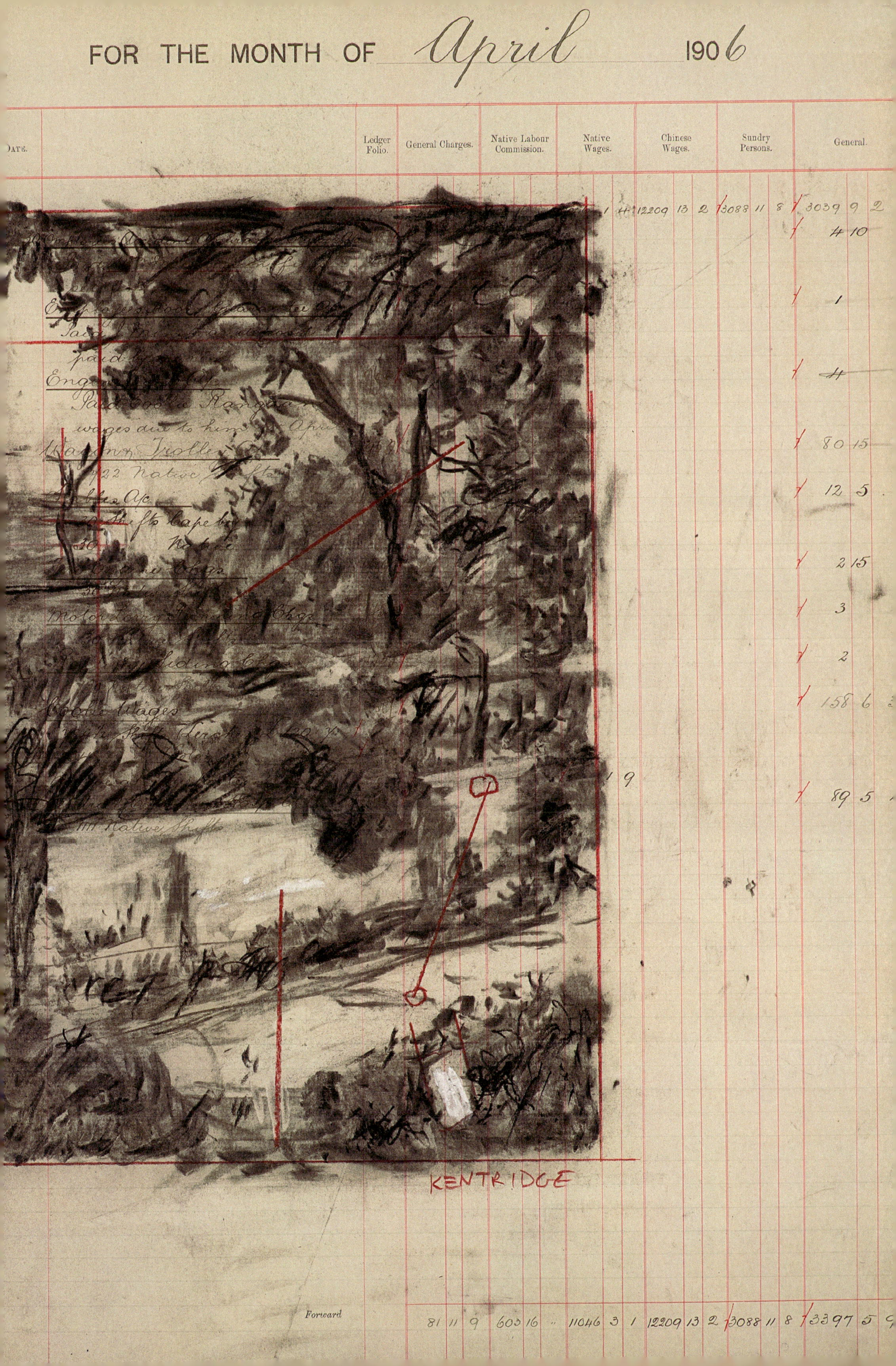
FOR THE MONTH OF April 1906
Date.
Ledger Folio.
General Charges.
Native Labour Commission.
Native Wages.
Chinese Wages.
Sundry Persons.
General.
KENTRIDGE
Forward

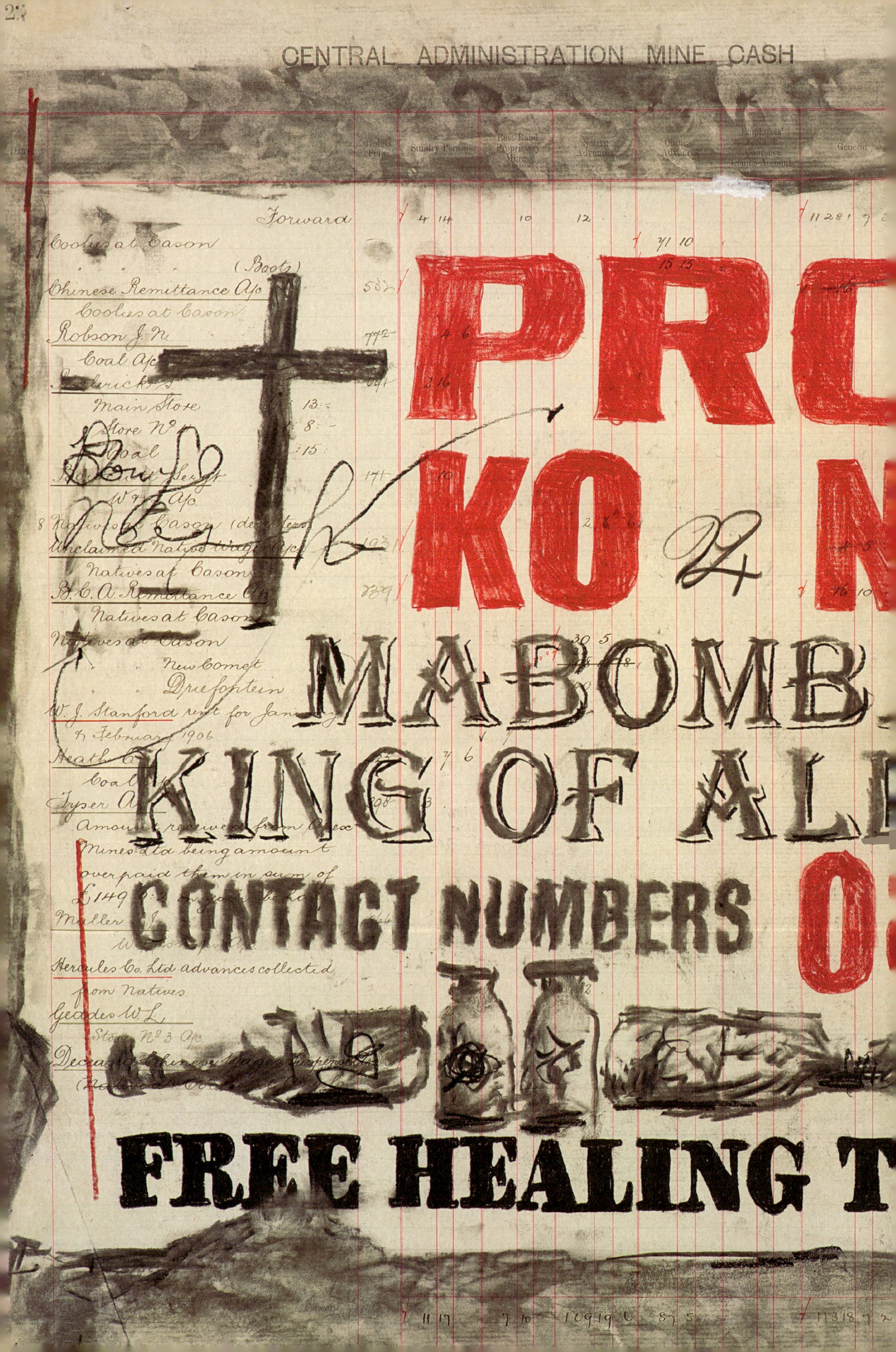
CENTRAL ADMINISTRATION MINE CASH
PRO
KO N
MABOMB
KING OF AL
CONTACT NUMBERS
FRFE HEALING T

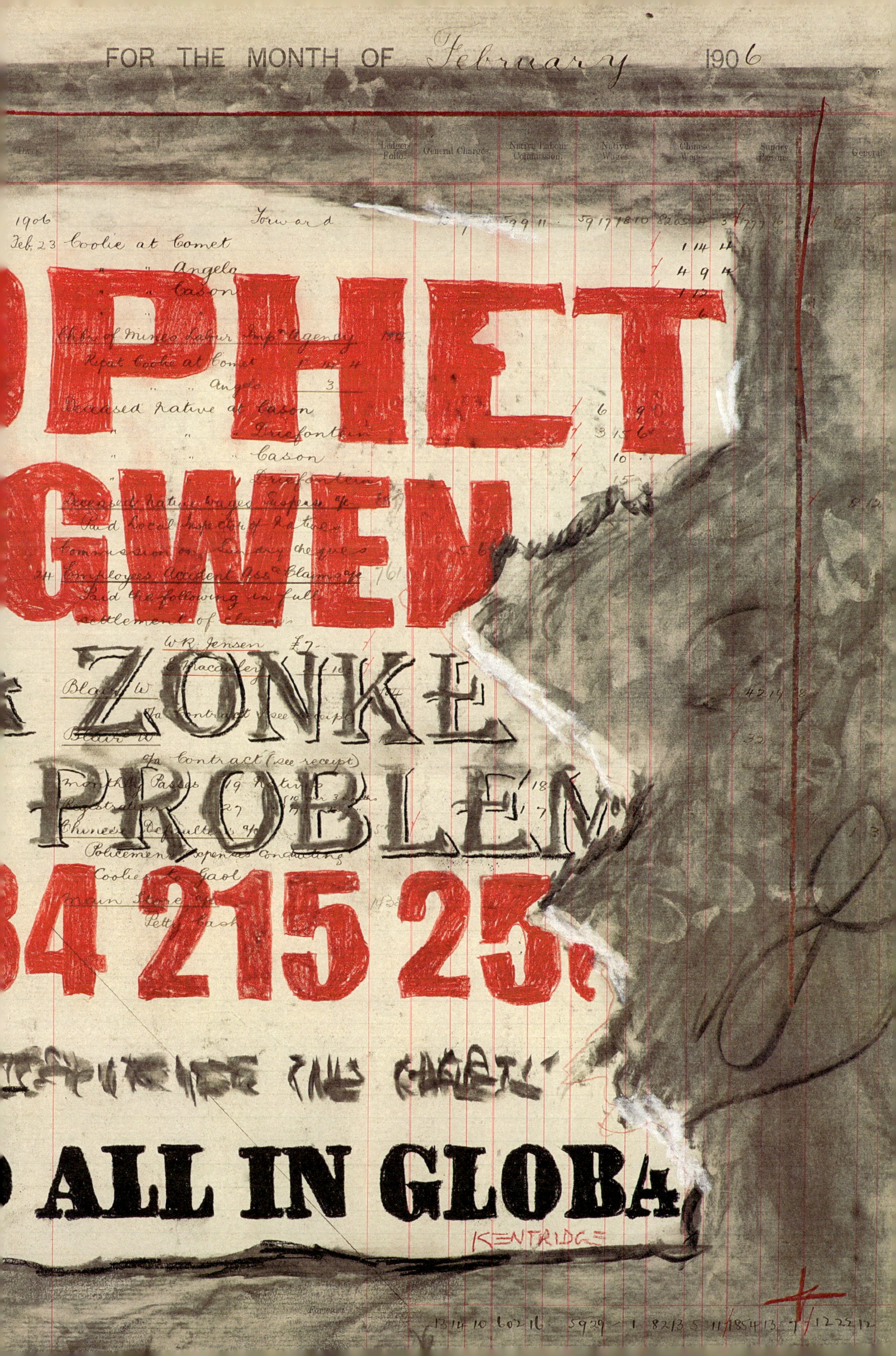
FOR THE MONTH OF February 1906
PHET
GWEN
ZONKE
PROBLEM
34 215 25
ALL IN GLOBA
KENTRIDGE

CENTRAL ADMINISTRATION MINE CASH

Date.		Ledger Folio.	Sundry Persons.	East Rand Proprietary Mines.	Native Advances.	Chinese Advances.	Employees' Accident Assurance Claims Account.	General.
906	Forward		471 17 3	796 2 10	220 11 4	1044 6 -	112 2 5	21848 5 1
23	Thorup J. E.	900	5 -					
	W & T a/c							
	Prior W	727	5 -					
	W & T a/c							
	Kooyker L	511	5 -					
	W & T a/c							
	A.R. Meaker rent of house for							
	November December & January			9 10 -				
	Levy S	541	3 7 6					
	Domestic Water a/c							
	Chbr of Mines Labour Impn Agency	154	1 17 4					
	Repat. Coolie at Comet 1. 14. 4							
	" " " Angelo 3							
	Coolies at Angelo					1 9 4		
	" " Cason					1 12 -		
	" " "					6 -		
	Natives at Cason				2 17 -			
	Deceased Native Wages Suspense a/c	689						6 19 3
	Decd Native at Cason 3. 3. 9							
	" " Driefontein 3. 15. 6							
	Native at Cason (decd)				10			
	" " Driefontein "				15			
	Deceased Native Wages Suspense a/c (nil)							
	Deceased Native Suspense a/c	687						16 10 -
	Received from Boksburg Gold Mines							
	wages due in the estate of deceased							
	native "Fifteen" Mine No. 167. Reg No 39349 £1-10-1							
	money found on body of "Fifteen" 15-							
	Hercules Co. advances collected							
	from natives				10 -			
	Chinese Capitation Fees	363						6 10 -
	Chinese Passage Fees	369						5 -
	Chinese Repatriation Fees	561						6 -
	Chinese Advances	475						15 9
	being balance of amount collected from							
	Coolie Jong Te Jim 5/9 making							
	total of £23-15-10 (see folio 6) and							
	amount collected from Coolie							
	"Regd" No. 2420 £18. purchasing							
	his discharge.							
24	Contractors Location a/c	842	10 10 -					
	Amounts collected from							
	natives at Angelo Location							
	Forward		488 7 1	805 12 10	234 13 4	1047 13 4	112 2 5	21890 - 1

FOR THE MONTH OF February 1906

Date.		Ledger Folio.	General Charges.	Native Labour Commission.	Native Wages.	Chinese Wages.	Sundry Persons.	General.
	Forward		2 19 8	584 7 –			39 6 6	353 14 5
7	Natives at D & A.				249 – 3			
	" Driefontein				2 1 –			
	Coolies " Cason					593 8 –		
	Driefontein Consolidated Mines Ltd.	587						1 17 –
	G. W. Solomons expenses to town with gold							
	Angelo Gold Mines Ltd	15						2 3 –
	G. W. Solomon's expenses to town with gold							
	New Comet G. M. Co Ltd	1560						1 14 –
	G. W. Solomons expenses to town with gold.							
8	Natives at Cason (Deserters)				6 5 6			
	B. C. A. Natives at Cason				26 1 9			
	Natives at Cason				518 9 2			
	" " New Comet				298 12 8			
	" " Driefontein				1889 7 6			
	Tyser A.	898					150	
	On a/c of Contract							
	Passes for 13 Natives			1 6 –				
	Regn " 15 " 7/8 over 4/3 mos "			15 –				
	Chinese Defaulters A/c	573						1 6 –
	Expenses incurred conducting Coolie deserters back to Mine							
	S. R. Haines expenses to Johannesburg with Natives			1 9 9				
	Chinese Defaulters A/c	573						2 – –
	Expenses incurred in connection with recovery of Coolie deserters							
	Employees Accident Assn. Claims A/c	761						9 10 –
	Paid J. W. Angwin in full settlement							
	Deceased Native Wages Suspense	689						17 17 2
	Paid Local Inspector of Natives							
9	Blair W.	104					100	
	On a/c of Contract							
	Cost of Colonial Secretary Pretoria re Hospital accommodation		– 5 3					
	Main Store A/c	1434						200
	Further on a/c of Petty Cash for January							
	Stamps for receipts		1 – –					
	Deceased Coolie at Angelo as per Pay Sheet					– 5 –		
	Forward		4 4 11	587 17 9	2989 17 10	593 13 –	289 6 6	590 1 7

CENTRAL ADMINISTRATION MINE CASH

Date.		Ledger Folio.	Sundry Persons.	East Rand Proprietary Mines.	Native Advances.	Chinese Advances.	Employees' Accident Assurance Claims Account.	Gener
	Forward							

ACCOUNTS

ROSALIND C. MORRIS

Before me sits a large book, bound in flocked caramel-coloured leather, the title embossed in gold and set into a red rectangle surrounded by tooled scroll that betrays the book's origins in an Italian craftsman's shop, some time towards the end of the nineteenth century. The title in the red rectangle reads: 'East Rand Proprietary Mines, Central Administration. Mine Cash'. There is a neat line (also gold) between 'Central Administration' and 'Mine Cash'. A horizon of sorts, this line is also a miniature glyph that evokes the dream-image of gold mining: above the horizon is the corporation, below lies the wealth that sustains it.

Or so the myth of mining would have us believe. But, as every miner knows, the wealth underground is not simply brought into the light. It is made—by working *in* the earth, though not *on* the land, despite the metaphors that can make mining seem like a mere reaping of the underground's fruit. That labour is more often than not invisible, ironically hidden in the bright penumbra of the gold bar, or the dazzling glitter of a necklace, the aura of a royal sceptre and crown. The question is: How does one illuminate what can be so easily secreted in the light of day? How does one look for shadow, to read the trace?

Open the book. Between the covers, which I recall with the aid of a digital image, and which is now before the reader in a persuasive facsimile, the paper is robust, though the years have bathed it in a soft sepia wash. The paper was of a very high quality, clearly, to have withstood the elements, the mould and the mites. But the book must also have been tenderly cared for and safely stowed during the long century after it was written.

Written. Already I hesitate. One cannot really say that this is a written book. It would be more accurate to say that it is a book that has been written upon. Or at least, one must admit that this book was not written in any usual sense. According to the title, this is an accounts register of the East Rand Proprietary Mines (ERPM) company, one of the largest and most significant mining finance houses in South Africa at the time of the gold industry's emergence. To say that it is written upon (and not written) is to recognize the strange kind of writing that comprises it. An accounts register is not a fiction, although what gets written there can make or at least influence a world. Nor is it a historical narrative, though one can infer the events of a bygone era from the explanatory notations that accompany every entry. It has a structure and one can discern the elements of grammar and syntax in the explanatory 'asides.' But they are of an uncommon sort.

Verso: Shaking Tables

Date		Ledger Folio	Sundry Persons	East Rand Proprietary Mines	Native Advances	Chinese Advances	Employees' Accident Assurance Claims Accounts	General

The structure of the register is like a scaffolding or a skeleton within the page itself—a latticework of red and blue lines making rows and columns that enable the double-entry account. What is written there both depends upon and floats above that structure. Notations of cost and expenditure are inscribed with remarkable detail, in a variety of inks—more and less blue, brown or black—by a small number of different hands. You can recognize the fluid cursive lines of one writer, the scratching haste of another; they are the signature marks of frontier functionaries whose names we have now forgotten. But these marks are also the traces of a fastidious, legally self-conscious writing machine, an assemblage of many people, many lines of authority, and many structures of decision-making. Beyond the matter of who made these notations, who double-checked them and marked the gesture with an under-lining, beyond who made corrections (in red pencil) or approved them (with a little blue tick mark), there is the question of what required them to be written. If there is something like an author behind the Cash Book, it is surely mining capital itself—of which East Rand Proprietary Mines is a fine example. One gets a fascinating picture of that impersonal writing machine from the Cash Book, although it is more like a negative than a printed image. It needs processing.

Recto

Date		Ledger Folio	General Charges	Native Labour Commis-sion	Native Wages	Chinese Wages	Sundry Persons	General

In general, I read the writing on this book and William Kentridge draws the writing on this book. Or rather, he draws on the book, the pages disassembled for that purpose. The 'Cash Book', as we have called it, was purchased by Kentridge from a private bookseller in Johannesburg. He found it in the dust-laden labyrinth where he often finds books for use in his artwork. When he showed it to me, my reaction and my desire—as a scholar who has conducted research on gold mining—was to preserve the text, or at least the understanding that might be derived from holding the pages together and reading them in sequence. So, we decided to photograph the book for archival purposes and, then, to work on it, in our respective manners and media, and to do so autonomously, before reassembling the pages we were re-imagining and re-imaging. In the end, we would produce our own book, the one before you. It would be a new object derived from the old one. It would be neither reducible to the previous book, the Cash Book, nor thinkable without reference to that original text. It would be at once a crypt and an epitaph—for both the Cash Book and the lives ghosted there.

Before we commenced our writing and drawing upon the book, we had to encounter the landscape to which it referred. But even there, travelling together, it seems we took different paths. Our division of imaginative labour turned out to be a division of spatial domains, or directionalities, and, perhaps too, of temporalities. William followed the horizontal; I the vertical. We drove across the land outside Johannesburg, stopping at random intervals, while he made visual notes and took photographs of the surface world. Not the space of surface rights referred to in the Cash Book, but the space of landscape.

Landscape has its own history of course, an art history. If the term has origins in early Dutch, the word *landschap* originally meaning region or tract, the aesthetic tradition of landscape has a modern provenance inseparable from classicist nostalgia and imperial romanticism. These are forces that resonate in the deepest recesses of the colonial consciousness of South Africa and in the unconscious of gold mining even today. They are referenced in some of the drawings that Kentridge has made on the pages of the Cash Book, just as they are summoned in the pages that follow here, in the words of poets and the geologists who quoted them. The Cash Book is our intertext. Nonetheless, the landscapes that emerge for Kentridge do so by virtue of associations and thought processes only tenuously bound to the words that dissolve beneath his charcoal and become what they are when meaning is exhausted—lines, marks, traces. The page is his medium, as it is mine, but different spirits are conjured there. His is the order of the visible; mine the sayable. And at the same time, his is the order of the invisible; mine the unsaid.

Mine the unsaid. That is my task here—to read from the meticulous but oblique references and calculations on these pages, the broader histories and narratives that make them possible. To discern the latter, one must both attend to the details that are written there, and, at the same time, bring to bear other archives and other stories. In this sense, the Cash Book has to be thought of not only as an archival repository but also as an echo chamber.

DATE: Enter Here

Where or when to begin?

The year 1906.

The Cash Book commences in February 1906, the balance from January 'carried forward'. For unknown reasons, it contains accounts for only alternate months. Somewhere, we surmise, there is another Cash Book, 'Number 1', and the rest of 1906: January, March, May, July, September and November. Perhaps the division of the Cash Book into two books, numbers one and two, protected the company against the loss of records in the event of fire or some other catastrophe. Perhaps, there are also other copies elsewhere, in which all the months are integrated. Perhaps the different registers were kept by different clerks as a check against mendacity, subterfuge and the sleight of accounting hand which makes of the accountant a thief. We don't know. But the idea that there might be, somewhere, another place where months are kept provides a compelling allegory for the yearning that must have afflicted those many migrant labourers who are counted and accounted in this often opaque register. For them, the memory of those places left behind and the dream of those to whom they might return at contract's end must have been as if in another, parallel time.

In 1906, South Africa became the world's leading producer of gold, a rank it held for more than a century, until 2008. In

1906, it produced 5,559,534 ounces of gold, valued at £23,615,400 (adjusted for inflation, that would be valued at £9,89 billion in 2010). 1906 was, nonetheless, the beginning of the third great depression of the young industrial economy, and one that threatened the entire fabric of what was, following the Second Anglo-Boer War of 1899–1902, a fragile emergent nation. The Transvaal, the site of the Witwatersrand and the gold mines, was then still a colony, having been annexed by Britain in 1900, though it would become a province within the new Union of South Africa in 1910. In the Cash Book, it appears in the references to the 'Prisons Department' of 'The Government of Transvaal', which supplied convict labour to the mines and recompensed warders, and 'overseers' but not the prisoners they supervised (p. 4r).*

* The verso and recto of each spread of the original Cash Book are henceforth referred to with *v* and *r* respectively.

Reading an archival document requires that one constantly shift focus. At first, it is the page that comes into view. Then, one focuses on a word that will not relinquish its meaning, its malformed letters written in haste, smudged or sullied by a splash of ink or coffee. Eventually, the script disappears and sense emerges from its stubbornly material medium. But then, after the estrangement of language, the long histories of use, etymological transformation and mishearing seem to clamour for recognition. As when one contemplates the word 'oversight'.

By virtue of a strange inversion within the English language, the word 'oversee' quickly becomes its opposite. Oversight can imply a failure of discernment as much as scrupulous supervision. And, in retrospect, every crisis seems to have been a matter of oversight, a failure to see. Each of the depressions (1890–91, 1896–97 and 1906–07) in South Africa was dominated by a particular crisis in the mining sector: the first a crisis of technology, the second a crisis of capital, and the third a crisis of labour linked partly to a crisis in technology. Each had a political element, and each was also related to developments and processes in other parts of the world. Complicating the picture is the fact that each of these crises appeared, initially, in the guise of another.

Reading and rereading the Cash Book allows one to recover some of the elements that have otherwise vanished in the mythic history of the gold mines as the foundation of the nation, and to thereby escape the illusions in which crises attired themselves.

What was common sense at the time now seems mysterious or improbable, or simply forgettable. So we must rediscover the ordinary, to borrow a phrase from Njabulo Ndebele.[1] It is a task not unlike that of rediscovering the comet named Holmes, which, having been discovered in 1892, just before ERPM was established, vanished from earthly vision in the year of the Cash Book, 1906. Comet was the name of an ERPM mine, and like all comets, it enters and departs the horizon of visibility only periodically, leaving mere dust in its wake. But, then, dust is the stuff of history and the archivist's purest element.

1 Njabulo Ndebele, *South African Literature and Culture: Rediscovery of the Ordinary* (Manchester: Manchester University Press, 1994 [1991]).

Comet Crusher: 2 Pounds, 10 Shillings

In 1906, according to the largest circulating newspaper in the region, the *Rand Daily Mail*, ERPM was the eighth most profitable mining finance corporation in the Transvaal. But the signs of potential catastrophe were visible as early as January—not only at ERPM but across the whole gold-mining sector. At the beginning of the year, 58 of 59 of the licensed mining companies registered a profit, but 1906 also saw an enormous loss of profitability relative to previous years.[2] The company suffering the worst diminution in profitability was East Rand Proprietary Mines. According to the wisdom of the press, it was simply not able to keep up with its own digging, its crushing facilities being too limited to convert the mountains of ore that were already being drawn from very deep levels. That ore had relatively low concentrations of gold, a fact that has overshadowed every aspect of the long history of mining in South Africa and that has been used to justify low wages as much as to incite new technological inventions.

2 *Rand Daily Mail*, 3 March 1906.

The Cash Book records the payment of bonuses to the overseers of crushers in February (p. 4r). No doubt, these sums are recognition for the additional labour that was necessary to keep up with the rising demands for ore processing and the limited facilities observed in the *Rand Daily Mail*. But, the fact that this 'overtime' was performed by convicts whose overseers were prison warders suggests how desperate the ERPM was to keep its labour costs down. For the bonuses are handsome indeed: £4 and £2.10 respectively, to each of the warders overseeing the facility bearing the irresistible name of 'Comet Crusher' in the accountant's abbreviated notation. The mines desired the convicts; had crushes on them, one might say. And no wonder. For, gold mining was and still is the realm of desire in excess.

The whole array of 59 companies lay cheek by jowl on a small wafer of land. In South Africa, gold is mainly to be found in an area called the Witwatersrand, a small stretch of sedimentary layers that sits upon the map in an elliptical shape stretching east, north and south of Johannesburg. It is, today as it was then, home to the deepest mines in the world, some now reaching more than 4 kilometres below the surface, accessed in multistorey iron-mesh elevator cars that resemble the cages used for animals in a zoo. The labyrinths of tunnels and shafts of the impenetrably dark places to which the elevators deliver the workers are sometimes remarkably banal. They have the accoutrements of any industrial workspace: toilets, first-aid facilities, repair shops, engine rooms, electrical stations, supply depots. At other times, they are hellish. The temperatures can approach 38 degrees, as hot as a fevered human body, although the average rate of temperature increase per metre is lower in South Africa than almost anywhere else in the world.

The mines are often damp and vulnerable to explosive collapses and inundation by water. For there are rivers beneath and within the earth, just as there are caves: it is a folded landscape that lies beneath the surface, lit only by artifice, and inhabited by the dead. Those residents are both fossils of ancient creatures that once populated an inland sea formed by converging rivers, and the ghosts of miners lost in a century of accidents: men, and now women, whose bones mingle with the earth, who have become dust and spirit both. Beneath a surface where cattle yet graze and corn is still harvested, a world nonetheless traversed by fantastic networks of highways and webs of water pipes and power lines, is a phantasmagorical world, a world of surfeit and surplus generated in the forge of violent extraction.

One hundred years after the Cash Book was closed, a guide at the Blyvooreitzicht gold mine (on the West Rand), where I was doing research, informed me that, behind the wall of rock where we stood, 2 kilometres below the surface, there was the water pressure of two Hoover Dams. Did he know that Hoover was indeed there, a shadow figure in the history of the supply of labour from China to South Africa's mines? That discovery would take me years to make, though it is also summoned by the Cash Book, where Chinese labour is marked and remarked under the headings of 'Chinese Advances'.

China would overtake South Africa as the world's largest producer of gold in 2008, but, for more than a century, South Africa held that position. The initial discovery on the Witwatersrand (in 1886) of alluvial gold—that which is loose at the earth's surface and extractable without much processing—lured attention and adventurers to the area around the barely emergent city of Johannesburg in enormous numbers. This was because gold had already been established as an object of desire. Commencing in the middle of the nineteenth century, the discovery of substantial gold deposits around the world and the long-running depletion of silver mines everywhere had undermined silver's status as the privileged medium of monetary value, ensuring that future gold discoveries would become the sites of manic investment. And so it was. Displacing all of the rushes in the world, from California to Australia and New Zealand, and Alaska to China, the Witwatersrand became the scene of a delirious drive for fantastic wealth.

ERPM had been formed shortly after the discoveries of gold, in May of 1893, with an initial capital base of £650,000—more than £200 million when adjusted for inflation today, and more than many of the other mining finance companies being created at the time. When it commenced operations, its assets included a small cluster of mining properties, including New Blue Sky, New Comet, Cinderella, Agnes Munro, Driefontein, St Angelo and Leeuwport. Its board of directors was chaired by George Farrar and included C. S. Goldmann, Lionel Phillips, J. C. A. Henderson and S. W. Jameson—giants in the history of South Africa's mining industry. Within two years of its formation, such was the promise of wealth on the Rand and the financial acumen of its board, ERPM had raised an additional £750,000 and commenced the restructuring operations that would allow it to enter the capital-intensive field of deep mining.

The names of the mines that appear in the Cash Book are those described as being of primary interest in 1903, when Goldmann, member of the ERPM board, published his monumental three-volume survey *South Africa Mines; Their Positions, Results and Developments*. Comet, Angelo and Driefontein and Cason mines are the most frequently named sites of ERPM's many operations and transactions in the Cash Book, the places to which 'Coolies' are sent and from which 'Natives' are returned from jail, at which accidents are noted and where funeral expenses are calculated, from which trips are made to the city and requests for new machinery sent. But there is also mention of Comet, Cinderella, Hercules and others. They are the sites to which orders of mealies and corn beer, coal and dynamite are sent, where workers claim their wages or petition for exemption from contracts and repatriation to their countries of origin—or sometimes, and against all odds, decide to stay and make a place.

A Blank Space: Between Time (Date) and Being (Sundry Persons)

In the columns that structure the account book, all but one are given a title. On both the recto (devoted to wages) and the verso pages (where advances are noted), there is a blank space between the date and the columns headed 'ledger folio' and 'sundry persons'. It is the largest of all columns, a blank space wherein narrative descriptions of the entries are made, the names of payees listed, the explanation for costs recounted. I think of these columns as the asides of the Cash Book, the places at which the writer paused, addressed himself or some other imagined reader and allowed himself to speak aloud, as it were, not only about the reality of what he was inscribing but also about the circumstances of that inscription. One doesn't normally expect much more than the most perfunctory notations in the columns of account books. And yet, there are three occasions in this 1906 Cash Book when, quite against the expectations of the genre, the accountant addresses someone quite directly, in the bureaucrat's version of the 'dear reader' more familiar from eighteenth-century novels. One could say that 'you' is there, along with all of 'them'. Erupting from the flat idiom of mere numbers, the accountant notes that £3 were paid to one 'A Tyser'. Registered as a value paid under sundry persons, it is described as the 'amount received from Apex Mines Ltd., being amount overpaid them in sum of £149:9:2 on *your* behalf' (p. 2v). Some months later, in June, under wages paid, also to sundry persons, there is another notation: 'Paid G. Blair Hook amount *due to you* as per C.B. No.1 fol. 36' (p. 33r).

And again, in August, there is another such reference: a 'reward to Ellen Richards for finding cheque *in your favour*' is entered under the name of Fisher Brothers.

In Your Favour, 'You' Is There

It seems likely that the accountant had someone present, someone to whom he was issuing these payments, and whose approval and consent to the truth of his notations was being solicited. We can imagine them, standing in a modest office, on a warm Highveld autumn day, the sky blue, perhaps, or, if late afternoon, clouded by storms. The book is there between them, with a bottle of ink, and, perhaps, on the wall above, an image of King Edward or a drawing of the Johannesburg cityscape, perhaps a list of things to be purchased in the city when the next messenger was sent. Some miners actually hanged X-ray images of their lungs on the walls next to torn pages from fashion magazines, the accumulating spots of the lung disease that would inevitably fell them matching the polka dots on the pin-up girls' dresses.

Whatever the tableau in which they were made, these entries comprise an extraordinary fold in the text, a kind of reflexive turn and verification that reaches beyond the page. The writer speaks to the recipient of funds, which must be explained. The recipient is asked to agree to the explanation. We can imagine a word of assent, a nod, a handshake at the door. But we cannot help but wonder if there is also another addressee of the book, George Farrar himself, the chairman of the board, founder of the company, and the voice haunting the machine of the Cash Book. Is he not there, in every 'you' written or unwritten in the text? A little god, if not the ghost in the machine?

In fact, there are only two overt references to the company's chairman in the entire register and they both confirm his remoteness from the operations of daily life at the mines that he owned and controlled. In the first, from 25 June, costs of 6 shillings, 3 pence are registered for a messenger, one F. A. Dowd, for taking Native Labour Reports to town 'for Sir George Farrar' (p. 26r). The second, on June 30, appears in the reported 'cost of telephone communication to Pretoria *to speak to* Sir George Farrar' (August, p. 41r). The telephone call is accounted at a steep 2 shillings, 6 pence (more than £40 by today's reckoning) but it is the honorific as much as the expense that leaps from the page. The knighthood had been bestowed on Farrar only in 1902, and with his baronetcy in 1911, he entered the group that would be referred to as the Randlords, the mining magnates who were recognized by the British crown following its victory in the Second Anglo-Boer War.

It was perhaps the power of the man or the idea of him that made the use of the new technology and its fantastical powers seem appropriate. There are many references to telegrams sent in the Cash Book, but only one to a telephone communication. In its careful notation of the fact of speaking, we can ourselves overhear what the technological magic of the telephone could enable: a sense of directly communicating with those who are absent. A strange energy must have been felt to course through the lines on that day, a special day, the last of the fiscal quarter. For telephones were still rarities in the Transvaal. One wonders what it sounded like: not just the voice but the line itself, crackling and full of gaps, the lags and blank spaces between utterances marking the distance between Pretoria and what is now Boksburg, a distance travelled today in about 45 minutes by car but requiring a good day by carriage back then.

Train Fares for Coolies

From 1890 onward, the mines along the line joining Krugersdorp and Springs had been linked by a railway system that also fed the trunk lines to the coastal port cities. Migrant labourers were sometimes transported in those cars; the second page of the Cash Book notes the costs of 'Train fares for Coolies' (p. 1r). But much of the local transport was via horse-drawn and ox-drawn carriages and, in the city, animal-powered trams, a concession for which had been granted to the Afrikaner-led City and Suburban Tramway Company in 1889. Originally founded by A. H. Nellmapius, it was soon floated on the stock market, and a majority of the shares were bought by the Rand Mining Company. The Tramway Company had been a major source of secure employment for Afrikaners, and a vehicle for wealthy agriculturalists to ensure a market for forage as well as a means of accessing the city's consumers of foodstuffs—but it was also a site of investment for the diversifying mining capitalists, whose economic power was in no way limited to the mines. In

1906, the first electrical tramway system opened in Johannesburg, forever changing the landscape and ensuring the further destitution of the stock value of the Tramway Company, as well as the loss of employment for many who had previously driven cabs—yet another factor in the depression of 1906.[3]

3 Charles Van Onselen, *New Babylon, New Nineveh* (Johannesburg: Jonathan Ball, 1982), pp.165–202.

Johannesburg may have glittered—or stunk—on the horizon, the glint of steel on the trams catching Highveld sun. On the mines in 1906, however, old-fashioned animal-powered carts were a living part of the machinery, and although an application was made in February for surface rights to extend the railway into ERPM's mines (p. 2v), one is not surprised to find, amid the many entries in the Cash Book, a note of monies paid to a Mr Penlerick, for oats (18 shillings) and forage (7 shillings, 6 pence) (p. 40v). Another entry refers to yokes and chains, a plough and a harrow (p. 36v). But even in a time of drought—and there was indeed drought in 1906—not all food is eaten. Across the page from the entry on oats, on the recto, is another reference, this one to bread—for use as an eraser to clean drawings (p. 41r). Although rubber erasers had been invented in 1770, the much-less-expensive crustless bread was still used to tidy the illustrations and maps on which the lines linking ERPM's properties to each other and the City of Johannesburg would appear as thin and sometimes dotted lines.

Surface Rights

If one lays the maps made by surveyors on the actual surface of the ground, they wrinkle and bend; the names and lines on the page often vanish in shadow. On lifting the map above the earth, each rigid line rises almost magically from the undulating horizon and becomes something wholly other than the earth it signifies. It becomes evidence of itself, of the coding of a landscape as measurable property, of the territorializing of power.

Unlike the landscape, the page is a flat space, and so it is the proper residence of a space flattened. Geometry, the grid as assemblage of line, square and rectangle, inscribes land-as-property, the form of the claim, the staked space. In mining, as in no other field, the prospector says, I claim what is mine. And the speculator knows, the mine depends on that claim.

In South Africa, as in most colonial contexts, the mapping, measurement and fencing of territory constituted a crucial gesture in the process of expropriating land from those who previously used it, and who lived with it in a different manner: mainly for grazing cattle and growing corn. Isabel Hofmeyr has rightly described this fencing as a kind of writing on the land, a process of coding it in the language of the bureaucratic state.[4] For it was only in being so coded that it could be divided, carved into territories that were transferred to settlers and mining claimants, and then differentiated between surface and subterranean domains.

4 Isabel Hofmeyr, *We Spend Our Years as a Tale That Is Told: Oral Historical Narrative in a South African Chiefdom* (Johannesburg: University of Witwatersrand Press, 1994), especially pp. 59–77.

There are many notations in the Cash Book about the purchase of surface rights, so that the activities below ground could be undertaken. In August, for example, such rights were sought to enable the construction of a court house exclusively dedicated to the trial of Chinese workers, a boarding house and keeper's cottage, staff quarters and an office for the mine manager, even a fence to protect the electrical station (p. 45r). They appear in the Cash Book secondarily, not as the licensing or other legal fees but as the cost of stamps on the applications.

The largest surface territory was devoted to crushing plants, tailings dumps and living quarters. Shaft heads and gear actually required little space. Indeed, at the time of a mines' establishment, the relatively modest surface marks of the mines, like the outcrops, betrayed the enormity of the spaces that would be carved beneath. In time, of course, the visible world would also show the signs of the vast caverns being produced beneath the surface: great mountains of crushed ore and processed earth would rise as the inverted image of the underground, much like photographic negatives disclose, to reveal in ghostly reversal, the shapes of the inhabited world.

For a long time, these huge trapezoidal mounds healed into a kind of second nature, becoming the 'mountains' of the Highveld. Now, they too are being mined for residual gold. No longer second nature, the second-hand gold shines as brightly as the surface of the dream machine that once stood atop the largest such dump in Johannesburg: the Top Star Drive-in Cinema. It was taken down, to permit the reprocessing of the mine dump's soil in 2006, exactly 100 years after the Cash Book opened to account for the value of a Comet and other sparkling mines.

Gambling School

On the Rand, the division of space into tiny parcels for the purpose of prospecting was followed by their reassemblage into larger and larger units, as adjacent claims were bought up and territories reconsolidated by financially powerful companies like ERPM. In the early stages of the mines' development, diggers staked claims that gave them rights to prospect on spaces where the gold was relatively close to the surface. Thereafter, the geology of the land made itself known; the gold was embedded in rock in narrow seams, and these were in reefs that slanted downward at a steep but flattening angle. Very soon, 'prospecting' became indistinguishable from mining and entailed sinking shafts on diagonal and vertical axes, with claims being staked of sufficiently large areas to cover possible and probable points of intersection deep below the surface of the earth. Already, by the early 1890s, it was known that the reef descended southward of the outcrops and companies with sufficient resources were buying up large tracts of land to the south of the outcrop even before they began prospecting. Indeed, the history of mines on the Rand can be seen developing in relatively neat rows of outcrop, first level and second level deeps. If these rows are imagined as pure forms, as horizontal and vertical lines, then they can also be imagined as shadow images of the columns of the Cash Book.

By the turn of the century, mines of 1 kilometre deep were not uncommon. There were, however, numerous technological obstacles to mining on the Rand—both because of the depth and because the ore below 120 metres was stubbornly bound to conglomerate rock by pyrite crystals. In the first place, there was need for a new kind of explosive, one that could be used deep underground where the rock was harder and less amenable to dynamite. In 1874, the Nobel company patented its new gelatin explosive to remedy the problem, and by 1894 it was being manufactured in South Africa by the Anglo-German Nobel Trust. But the gold-lined ore thus liberated was not susceptible to the usual processes of crushing and amalgamation with mercury. For several years, there was no known technology capable of separating out the gold and it was during this period of waiting for invention that the first great depression in the mining economy occurred. The solution to the technological problem came from Scotland, where a cyanide-based leaching process was developed in 1887 by John Stewart MacArthur, with financing from the Forrest brothers. But the solution and the end of the depression entailed a Faustian bargain, insofar as cyanide is a highly toxic chemical that often remains in the effluent or tailings ponds and streams around gold mines, along with the lime, sodium hydroxide and lead nitrate that are used to prevent the cyanide from becoming even more poisonous. If these artificial lakes of turquoise with their beaches of gold sands resemble holiday resorts, as they sometimes do—despite the signage that warns children away with piratical skulls and cross-bones—it is but one more instance of the phantasmagoria of gold mining. Look, but don't touch.

Much of the capital needed for the new technology as well as land acquisition came from the increasingly monopolistic diamond-mining industry that, a few decades previously, had been established in Kimberley. But a lot would also come from Europe, from the stock exchanges in London and New York. Those distant investors were mainly speculators; and indeed, most of the companies established as joint-stock entities at the time would never be involved in mining at all. In fact, of the more than 500 companies listed on the market in 1889, only 44 were producing mining companies.[5]

5 Allen, *The History of Black Mineworkers*, p. 135.

Head Office, Paid in Full

Farrar was the only one of the Randlords who hadn't made his fortune in Kimberley and ERPM was in this sense relatively dependent on foreign capital. But the board structures were complex and the line dividing local from international capital was not an easy one to draw. The South African mining scene at the turn of the twentieth century was dominated by a few groups of capitalists who sometimes appeared to be fierce competitors of one another (especially when seeking or obtaining new stakes and when producing assays and other assessments of likely profitability). But they also collaborated with one another when it came to managing labour and seeking legislative interventions or other structural conditions that could ensure stable and cheap access to transportation, power, water and other necessities.

Historiography that concerns itself with the Anglo-Boer War focuses on the differences between these groups, emphasizing the fact that those most invested in deep-level mining were also most likely to have supported or been directly involved in the failed Jameson Raid, which sought to depose the Boer government of Paul Kruger and what was perceived as his industry-hostile policy in favour of an ostensibly mining-friendly British colonial administration. On the other side of the 'conspiracy' were those with investments in outcrop mining. Outcrop mining could function with relatively short-term horizons. It was less capital-intensive and it required little of the large-scale organizational structuring that deep-level mining demanded. For this reason, it is often said, its participants were less concerned with long-term governmental policies and were less likely to revolt against Kruger's government.

George Farrar, ERPM's chairman, was not only a 'conspirator' but was also initially sentenced to death for his leading part in the 'treasonous' assault. Kruger later pardoned him and he was subsequently restored to his place in the circles of power (and then given a knighthood!)—at least until 1911, when his reckless administration of his own company led to an internal coup by other directors of the board. Even so, that same year saw him granted a baronetcy. A soldier and cavalryman, Farrar was used to both conflict and compensation.

Company Accounts

Historiography that is less concerned with the war and more with the structure and social implications of the mining economy has a different story to tell: of more complex affiliations and overlapping, rather than competing investments.[6] The boards of directors of the mining companies in 1906 are a picture of collaboration more than they are a portrait of competition. One has only to look at Farrar's board memberships, as recorded in Goldmann's *South African Mines*, to grasp the extent of the lapidary network that saw a few men oversee and control the most lucrative mining industry in the world.

Farrar sat on the following boards as a director, sometimes as chairman of the board, sometimes as a significant shareholder: Agnes Munro Mining Company, Ltd, Angelo Mines, Ltd (Chairman), Apex Mines, Benoni Gold Mining and Exploration Syndicate, Ltd (Chairman), Benoni Gold Mines, Ltd, Chimes Deep Levels, Ltd, Cinderella Gold Mining Company, Ltd, Driefontein

6 Since the mid-1960s, when Geoffery Blainey called on historians to consider the material basis of mining in their analyses of everything from the Jameson Raid to the Anglo-Boer War and the Reconstruction period, most of the economic history of mining in South Africa has been concerned to understand how the move from outcrop to deep underground mining shaped the nature of the economy, enabling or summoning the complex structures of finance and industrial capital that would be the hallmark of the country's mining industry, and summoning one or another labour strategy. The contributors to this debate are numerous and the differences in their approaches too complex and subtle to be reproduced here. But readers wishing to pursue the matter should consult some of the following: G. Blainey, 'Lost Causes of the Jameson Raid', *Economic History Review*, 18(2) (1965): 350–66; D. T. J. Denoon, 'The Transvaal Labour Crisis, 1901–1906', *The Journal of African History* 8(3) (1967): 481–941, and 'Capital and Capitalists in the Transvaal in the 1890s and 1900s', *Historical Journal* 23(1) (1980): 111–32; Alan Jeeves, 'The Control of Migratory Labour on the South African Gold Mines in the Era of Kruger and Milner', *Journal of Southern African Studies*, 2(1) (1975): 3–29; Robert Kubicek, 'The Randlords of 1896: A Reassessment', *Journal of British Studies*, 11(2) (May 1972): 84–103; Shula Marks and Stanley Trapido, 'Lord Milner and the South African State', *History Workshop* 8 (Autumn 1979): 50–80; Martin Legassick, 'South Africa: Capital Accumulation and Violence', *Economy and Society* 3(3) (1974): 253–91; A. A. Mawbey, 'Capital, Government and Politics in the Transvaal, 1900–1907: A Revision and a Reversion', *Historical Journal* 17 (1974): 387–415; R. Mendelsohn, 'Blainey and the Jameson Raid: the Debate Renewed', Collected Seminar Papers, Institute of Commonwealth Studies 22, pp. 50–9; Peter Richardson and Jean-Jacques van Helton, 'The Development of the South African Gold Mining Industry, 1895–1918', *The Economic History Review* 37(3) (1984): 319–40; H. Wolpe, 'Capitalism and Cheap Labour Power in South Africa: From Segregation to *Apartheid*', *Economy and Society* 1(4) (1972): 425–56.

Consolidated Mines, Ltd (Chairman), Driefontein Gold Mining Company, Ltd, George and May Gold Mining Company, Ltd, G. F. Syndicate, Henry Nourse Gold Mining Company, Ltd, H. F. Syndicate, Kleinfontein Central Gold Mining Company, Langlaagte Block B Deep Level Gold Mining Company, May Consolidated Gold Mining Company, Ltd, New Blue Sky Gold Mining Company, New Comet Gold Mining Co., Ltd, New Kleinfontein Company, Ltd (Chairman), Rand Klipfontein Company, Ltd, St. Angelo Mines, Ltd, Simmer and Jack Gold Mining Company, Ltd., Simmer and Jack West, Ltd, The United Langlaagte Gold Mining Company, Ltd, Wolhuter Gold Mining Company, Ltd. The list boggles the mind. Not merely for the number of mines but also for their diversity—some are more concerned with outcrop mining, some with deep-level mining, some have boards dominated by one clique of capitalists, others by another.

The groups of South Africa's mining sector are generally described as the Werhner Beit complex, the Consolidated Gold Fields complex and the Farrar/Anglo-French complex. But Lionel Phillips, who was the partner of Friedrich G. J. Eckstein in Wehrner Beit, was a member of the Board of ERPM, and Farrar was on the board of the Nourse and Wolhuter companies, which were officially in the Werhner Beit–owned group of mines. If they were competitors, one might say, it was often a friendly match.

Convict Accounts: Cinderella, Hercules and a Saint (No Angel)

Other than Comet, the names of the mines within ERPM's orbit form a list of fairy tales and fables populated by little gods and heroes, saints and references to elsewhere. It is never possible to finally translate a name or to provide a true history of its origins, but the names of ERPM's mines also seem to reveal a great deal about the mythic unconscious within which mining was undertaken. Cinderella and Hercules seem particularly apt. One comes from a tale of fragile beauty hidden beneath coal dust (cinders); the other is a character that traverses classical myth as an icon of extraordinary strength, someone who can enter the underworld and return unscathed.

Many of the gold mines on the East Rand were located beneath collieries; coal covered the rock formations in which gold was to be found. They were all Cinderellas in that sense. But, as we know, Cinderella also had cruel siblings. In fact, the Cinderella mine drew much of its labour from the Cinderella Prison, which was established in 1905—just before the Cash Book was opened. This is why there are so many references to convict labour in the Cash Book and to the bonuses paid to wardens and other staff of the Prisons Department.

In his incomparable study of the early history of the Witwatersrand, Charles van Onselen makes clear that the prison was part of a tightly knit complex that included both concentration camps and mine compounds. They shared and developed a wide array of technologies for tracking people, including early biometric practices.[7] Those technologies are also registered in the Cash Book, accounted, for example, in the expenses for one A. P. Norton, who was sent to town on a matter related to 'finger impression recording' (p. 65r).

7 Van Onselen, *New Babylon, New Nineveh*, p. 377. See also Allen, *The History of Black Mineworkers*, p. 125.

In a world where prisoners become miners and miners prisoners, it is inevitable that law would breed its outlaws. And it should come as no surprise that it was from within the prison-mine-camp complex of ERPM's portfolio that the first millenialist criminal gangs, known as the Ninevites, formed around the spectacular and mysterious strong-man of Jan Note, aka Nongoloza. Van Onselen describes him in the idiom of local rumour, as the man with the piercing eyes; others perceived him as the phantom bearer of a terror that could erupt anywhere. After 1906, the Ninevites would develop strongholds in almost all of ERPM's mines, especially those named in the Cash Book—Angelo, Comet, Cason and Driefontein.[8]

8 Van Onselen, *New Babylon, New Nineveh*, pp. 375–8.

Received from Hercules

As for the other figureheads of ERPM's mines, Hercules was, of course, the Roman name for the Greek god, and he was said by Tacitus to be a favourite of the Germanic peoples. In a country where the claim to Roman ancestry, via its putative reinscription in Dutch-Roman law, dominated what was emerging as Afrikaner nationalism, he seems an inevitable mascot. One cannot help recall that one of the ten labours that proved Hercules' strength was the cleansing of the Augean stables in a single day (Augeus was said to have had the largest herd of cattle in the world). In South Africa, such a figure must surely have appeared (if only unconsciously) both to represent the ideal of pastoralist nations, like the Zulu and Xhosa, and to offer an image of colonial power after it had appropriated the cattle of those whom it conquered.

Such gestures of expropriation, which seeded the fields of African rage throughout the nineteenth century, were reported as late as March of 1906. As waves of Zulu protest against new legislation moved across Natal, the *Rand Daily Mail* reported the auctioning of livestock confiscated from insurgents in Mapumulo. Within a few months, ERPM would be compensating white miners who chose to leave the Transvaal to fight against other Zulu rebels under the leadership of Bambatha, whom the *Rand Daily Mail* referred to as the 'turbulent chief of Greytown'. They are noted under a new heading in the Cash Book, namely the 'Volunteer Compensation Account'. As for the insurgents, they are absent, if ironically intimated in the uncannily resonant term, 'Native Advances', which phrase heads the sixth column of the verso page.

Of Native Advances, much can and will be said, but we have not yet finished our survey of names. At the centre of ERPM's properties was 'St Angelo', renamed Angelo Gold Mine following financial restructuring. The Catholic provenance of the name is ambiguous, but it is tempting to surmise that St Angelo was an appealing totem because he (at least one of the saints who bore that name) had the capacity to be in two places at once. The definitive quality of the geology of the East Rand was, after all, the existence of two, and sometimes three, reefs at different depths. And the most successful mines were those that could operate at both surface and deep levels—in two places at once. To be in two places at once is what the hagiographers call the miracle of bilocation. But it is also what the mineworkers experienced as the predicament of cyclic labour migration, a predicament whose psychic manifestation is often termed double consciousness. These workers traversed the space between places, traveling from rural areas in Natal, Mozambique, Swaziland and elsewhere, some coming from as far as China. Compelled into wage labour by tax and other laws, we may nonetheless wonder: What brought them across sea and mountain, to the world of underground rivers and canyons?

Sundry Persons

There is something narcotic in the idea of gold, for those who would own it. The Witwatersrand gold rush was born, as the madness of gold always is born, from the fantasy that there is wealth to be harvested from the earth itself, wealth of a strange and magical sort. For gold is a useless metal—too soft to sustain industrial applications or to hold its form except when encrusting other, sturdier shapes. Its value in the late nineteenth century lay in this very uselessness, which made it, for some time, the perfect measure of all other values and as such, a commodity of its own inestimable value. For centuries, that very uselessness had made gold the perfect medium for adornment, the sign of surfeit. But beyond the spectacle of adornment, gold is the ultimate fetish, the medium and the sign of an economic logic that depends on the secular miracle of transubstantiation—of exchanging one thing for another, through an object that is both inside and outside of that relation. During the years of the gold standard, and even today, when people hoard gold as a hedge against volatile capital markets, gold was and is still the groundless ground of an economy whose hallucinatory violence depends on appearing as though there is, somewhere, a reserve, in which is stored the truth of value. This shared dream has been implanted in the world, sometimes as a reverie and sometimes as a nightmare. It all depends on where you are in the rush. The owners of the wealth never appear in the Cash Book, save in the person of George Farrar. But the workers do, beneath three distinct headings: Sundry Persons, Natives and Chinese.

In the Cash Book, sundry means European. Somewhat enigmatic from today's perspective, the word seems to contain the whole contradictory history of South Africa. Its primary meaning is 'diverse'. But its secondary meaning, derived from the Old and Middle English term, 'syndrig', conveys a slightly different nuance, namely 'separate' or 'private'. We hear it in the phrase, 'torn asunder'. In 1906, the formal concept of apartheid had not been invented and Afrikaans had not yet become the special medium of the discourse of separateness (apartheid) and the practice of separation. But that political philosophy is there, already, in the English—long before the rise of the National Party and its ideology of apartness and separate development.

Under the category of sundry persons one finds names of the European miners, many from Wales, Cornwall and Cumberland. In the early days, following the first discovery of gold, there were also miners from Australia and the United States, Canada and New Zealand, men who had cut their teeth on the rushes elsewhere and who hoped to strike it rich quickly. But all come together in the single column that holds those with recognizably European names. The first item on page 1 is, in fact, a list of unclaimed European Wages, apportioned as follows:

J. Stone	£18:14:6
J. Whitford	2:14:6
E. J. Willey	56: 9:6
E. Tippet	70: 7: –
A. Lawry	12: 5:3
R. H. Blyvelt	25: – :9

On the same page, there are coal accounts for Allan, Armstrong and Ambrose, and one for Morris. And Walker is listed as owing under the Main Store Account. Heating the little cottages rented to the white miners was a considerable expense for the miners, and there were few sources for supplies other than the company stores, one on each mine. So it is not surprising that the little stores, which extended credit until payday and then reaped the hard fruit of the sudden cash flow, are frequently noted in the Cash Book. Un-noted because always paid in full, the biggest purchases occurred as miners prepared to return home after the expiration of contracts. Then, as now, little carts were heaped with goods to take elsewhere. The hoarding of hard-earned coins, which, if not spent on beer and women, or, in some cases, laundry, were found in the pockets of dead men, acquired its purpose at last in these extravagant purchases that were the proof and the reward of survival.

Not everyone made it home, of course, and many bore scars and injuries that made future work impossible. A few pages into the Cash Book, the 'Rand Mutual Association Company' heads a list of more European workers, including:

Hosking E
Holman W
McCuster J

Allen W
Cullen W
Masters WJ
Hensen WR
Nottt G H
Knox S

Each name marks the spot of a worker, a consumer, a debtor and a creditor, and also, of a man wounded. It also marks the spot of a possible death and sometimes, in other columns, of a corpse, a place where the one who occupied that name no longer speaks. Suspended on the ladder of the Cash Book's scaffold, the same name holds these different moments of the white miners' lives within the space of a few letters.

Letters were sent, too—from the mines to homes around the world, with news of life on the Rand. In fact, 'letters home' were not only written by miners to distant mothers and wives; they constituted a veritable genre of both literary aspiration and newspaper reportage, giving those in small towns in Wales and China, Lithuania and Australia, cameos of life on the Highveld. The archivist finds in them a veritable 'gold mine' of detail about everyday life: reports about the weather and the food, the cost of ordinary goods and the style of housing, moral commentary and advice for the would-be immigrant or contract worker. So important was this epistolary relay that the first proposals to import Chinese workers to the mines included the provision of paper and stamps and address systems that would permit recognition of the destination by the postal services in both South Africa and China. They were deemed as essential as the provision of food and clothing. It is for this reason that on 4 October 1906, the postage for Chinese letters is noted in the Cash Book under the heading of the 'Chamber of Mines Labour Importing Agency' (p. 66r).

Miners from the European mining traditions—rather than the prospecting frenzies of the frontiers—were often from families in which generations had mined. They had a sense of vocation, were often militant, and their demands were sometimes met in anticipation by managements that feared union organizing. But their experiences on the mines were unlike those of the Native labourers whom they often exploited as both workers and servants, the latter sometimes bearing the stigma of the term 'boss-boy'. The tradition of boss-boys had some precedent in the diamond fields at Kimberley, where white diggers employed black assistant miners on short-term contracts. But the Highveld saw an entirely new development of this ironically privileged servitude. Along with male brides for unaccompanied miners, the Rand saw the institutionalization of the boss-boy. But he who was a bossed man, with his tea and sjambok, also ordered about his fellow Africans on behalf of his white patron. As such, he materialized a rift in the social world of African miners to match that of the earth, but it was only one element in the severe stratification of mining labour that would be intensified in a racialized form until the apartheid regime was overthrown, in no small part thanks to the resistance organized by black miners. Even today, mining remains one of the most hierarchical of labour sectors. There are no longer boss-boys but there are still bossed men.

Of course, there were actually boys on the mines in 1906, such as 'Picannin No. 48' who is noted in the Cash Book's June columns for having celebrated his fifteenth birthday on 1 May (p. 87r). But the institution of the boss-boy was something else, and it had already become a subject of concern in 1903 when Frederick Ingle testified to the Transvaal Labour Commission that white miners were making more in South Africa for less output than anywhere else in the world, and, in the process, 'wasting' black labour.[9] Ingle believed that the boss-boy was evidence of white lethargy and decadence but others recognized in it a strategy of divide and rule.

We don't know who among the many Native workers inhabited the word 'boss-boy'. What we have in the Cash Book is the mere trace of that order's nascence in the system of its nominations: the sundry persons, named one by one; and, in their shadow, the sundered people, those whom the Sotho singers of *sefela* would call the 'in-a-file people' in their incantatory narrations of their journeys to the mines.[10]

In Name Of

The black workers appear in the Cash Book as 'Natives'. Next to the Europeans, enumerated but also named as the renters of cottages, the holders of credit at the company stores and the recipients of insurance payments following injuries in mining

9 Allen, *The History of Black Mineworkers*, p. 147.

10 *Sefela* is a genre of narrative song sung by Sotho migrant labourers. It is often full of poetic detail about the mines and varies in tone from lamentation to sharp irony. See David Coplan, *In the Time of Cannibals: The Word Music of South Africa's Basotho Migrants* (Chicago: University of Chicago Press, 1994).

accidents, the Natives occupy a space of relative anonymity and non-recognition. Only a very few among the more than 10,000 Native workers accounted in the Cash Book are given their proper names, and for this very reason they elicit attention, sitting as they do at the still-opaque centre of a knotted and incomplete narrative. Thus, for example, the Native 'Mpali' has his fare covered to Flagstaff, where in all probability he was employed to recruit more Native Labour. We know this because his name is entered under the heading of the 'Chamber of Mines Labour Importing Agency'. There is also a Native named James, whose wages are paid (p. 7v), and a man named Diamond, alias Philip, whose insurance payout was recovered from the 'Coloured Insurance account' (p. 48v).

Another, named 'Campbell', has a ticket paid from Johannesburg to the East Rand. His earnings for April and June are noted, as is the payment of his daily expenses in October, at 1 shilling per day—unusual notations for a Native employee whose compatriots were almost always described only as a member of the lumpen category. What did Campbell do that drew him to such attention, made him the exception against which the other men descended into anonymity as a rule? It seems likely that Campbell was actually employed as a gardener by the ERPM, for he appears under the 'Gardening Account', just below the entry for A. Vlaskamps, whose expenses for a trip to town are paid in order that he be able to 'select Flower seeds' (p. 66r).

Select flower seeds. How odd is this little token of a bucolic world, of blossoms amid the drills and the dynamite. In fact, the tradition of the company gardens was already well established by 1906. The mining companies' flower beds were a citation of that first company garden established in Cape Town in 1652 by the residents of the Dutch East India Company. Then, it had been a garden wherein the fruits that would sustain the naval forces of the merchant empire could be nurtured. By 1906, however, the company gardens had become a mere ornament, no more intrinsic to corporate operations than the plumage on a Roman general's metallic helmet is to his military strategy. Even today, such gardens contain neat beds of roses, splashes of portulaca and pinks, and marigolds to keep the pests way. But in their orderly profusion of colour they are more reminiscent of the surveyor's mapped plots than the English country gardens, which they nonetheless aspire to recall. On the russet soil of the Highveld, a soil reminiscent of dried blood even beneath the golden glow of summer grasses, these little gardens were and still remain the tokens of mining's improbable claim to being part of the pastoral world. No management compound is complete without them. And almost every mining town has at its perimeter, where the highway meets the town, a signage of blossoms, a fragrant welcome to the unfamiliar visitor.

If I am correct, Campbell was a man who worked on the earth but not in or beneath it. Counterposed to him, an improbable figure of the pastoralist being in exile, are the ironically named mining 'Natives' from other nations. They appear as the 'East Coast Natives' under the heading of 'Natives Return Home Account', recruited or 'procured', as the language of the time had it, by the one who is named in the Cash Book as the Portuguese 'curator' (pp. 1r, 15r). These workers, largely from Mozambique, were recruited and supplied to mines by the Witwatersrand Native Labour Association and they accounted, on average, for nearly half the workers on the East Rand in the years between 1902 and 1907.

The oxymoronically named East Coast Natives are accounted only in numbers. But between the named man and the anonymous crowd of migrants, there is one called 'Fifteen', a Native who died at Mine No. 167. He is inscribed with both name and registration number, the latter being the number given him when he was enlisted by the Chamber of Mines Labour Recruiting Agency: No. 39349 (February, p. 8v). The miners emanating from China are noted in the Cash Book with two numbers, one given to them by the Foreign Labour Importing Agency and one that designates their pass number once registered with the Chamber of Mines.

How shall we reckon this record? In sum, what does it mean when a man bears a number as his name and is given a number in lieu of a name in the account book of his contractor? Fifteen, aka No. 39349, died with £1.10 on his body. The amount was duly entered and accounted in the Cash Book under the 'Deceased Natives Suspense Account'. But it is referred to as the wages due from the Boksburg Mine, where Fifteen was perhaps 'on loan' when he was caught in an explosion or lodged between heavy machinery or felled by loose rock—or, perhaps, mortally wounded in some other event born of personal malice or simple envy in the off-hours. We cannot discern the nature of the event that devastated him; there is nothing else in the Cash Book about the Native named Fifteen.

Fifteen. This name which is not one merely demonstrates that the Native is, by definition, the unnamed. Expelled from the language of power, he is the one who is not given an account, who doesn't count—even when he bears a pseudo-name (if not a pseudonym) in the image of the counted. But that only begs the question.

On Account of Contract

Who counts? Who does the counting? Who counts as one to whom one owes an account? Somewhere, someone has inked a finger, measured a naked body, weighed it, listened to its breathing, its heartbeat. Such was the practice of recruiters and labour brokers in the time of the Cash Book (and until recently) that the miner's body was weighed before and after it left the mine. When, after a few years of service in the underground, it began to slough off its muscle, starved by its spotted and shrinking lungs, it would be cut loose from the world of the contract, allowed to float, like discarded paper ruffling across the veld in the wind. And to die, wheezing.

In the Cash Book, one might say, 'Native' is a form of *name-calling*, but not properly an act of naming. The name is always a noun, but a common noun is given as a name only to those who will be buried in a common grave, those who are ghosts in the graveyard of the commons. Fifteen is noted in the Cash Book, it seems, only because he had a pound and 10 pence in his pocket. Death redeems the anonymous workers when money is owed, or found. Thus, the names of non-European labourers appear mainly in such cases when, after their deaths, their pockets turn out to be full. Or when, as we saw above, they become the exceptions to the rule in colonial capitalism's hothouse garden.

There are dozens of entries in the Cash Book under the heading of 'Deceased Natives Account' but generally only the pass numbers are given to identify the dead man. On rare occasions, the tally of shifts for which monies are owed to a worker who has transpired in the 'Central Coloured Hospital' is noted as well as the sums. Only once, in all the pages, do we get a list of Native names like that which appears so regularly in the column of sundry persons. It occurs on June 15, when payment is made to the Native commissioner in Johannesburg in the names of five workers, to whose surviving families the chamber was entrusted the task of transmitting unpaid wages (p. 37r).

101212	Matches	1:15:–
101250	Johnny	1:15:–
100372	Kelan	2:4:–
39924	Jim	2:4:–
48563	Rice	3:18:1
48563	Rice	:2:3

The dead men had worked for months on their contracts by the time of their deaths. In 1906, a three-month contract for black workers earned a little more than £1.5 and a year's contract slightly over £5.

Matches and Johnny had worked just less than three months, judging by the entries, Kelan and Jim for close to six months, Rice nearly a year. Just what caused their deaths remains unexplained in the Cash Book. And why they, among the thousands of Natives whose passes are paid for in bulk, many of whom also died, are given names is something of a mystery. Was it the enormity or horror of a particular accident that singled them out, in death, for a recognition denied in life?

Return Home

The Native goes without a signature in this world, where the only indices are the accumulating scars. But how did the Native become someone who could be, as is noted in the Cash Book, 'repatriated'? On 7 April 1906, repatriation of Natives from Driefontein mine cost 31 pounds, 12 shillings and 4 pence. In the same column, a later notation refers to the railway fares for 'boys returning home'. About half way through the Cash Book, the term 'Native Return Home Account' appears as a heading on the verso pages. To what nation does the Native go? Whence does he come, enabled by the passes that are counted here, in these pages, by the thousands?

As is well known, to become a Native in South Africa was to be evicted from space. Shunted into the margins of increasingly unviable land, into 'locations' where neither grazing nor the growing of maize could be done on any significant level, Africans were then forced into migrant labour by legislation that taxed individuals and huts, and demanded payment in cash. The sequence of legislation that began in 1869 and that was elaborated in seven subsequent acts prior to 1899, tightened the net around the locations and made it compulsory for all able-bodied men to pay a poll tax.

Marx had called this process 'originary accumulation' (it is sometimes translated as primitive accumulation) and like all originary processes, it would have to be constantly repeated, the invitation to urbanity (or the pseudo-urbanity of the industrial mining town) renewed and withdrawn, the demand for movement inscribed in the closed circle of migrancy, and always followed by a commandment to leave. Nor has this changed, although the coercion is now purely economic. Today, despite the settling of mining townships and the transformation of the polity, a large proportion of miners continue to make seasonal trips from the country's destitute periphery to work in the mines. They live in hostels or informal settlements, often maintaining two separate households and returning periodically to live in the simulacral image of rural tradition. I say simulacral because these worlds are so much the product of industrialization and were so artificially generated by colonial and apartheid land legislation, that the idea of tradition must be read as a sign of the new—albeit a newness that is itself now more than a century old.

Of all the figures of that tradition, it may be that 'kaffir beer' remains the most seductive. Distilled alcohol, which had been the monopoly of the Afrikaners, and then a major investment of the mining companies themselves, was prohibited when the Chamber of Mines realized that what had initially been a lure for the Mozambican miners had become a burden to the mines that had promoted it for this purpose. For drunkenness meant less productivity and more accidents. The role of alcohol in the transformation of South Africa was nonetheless recognized in 1906 with a certain wistfulness when the former chairman of the Witwatersrand Labour Association claimed that drink had made the Mozambicans the core of a new industrial society—because the debts occasioned by alcohol could only be repaid with money earned on the mines.[11] And if prohibition was the motto of a new order of efficiency, kaffir beer was to become the traditional alibi for keeping the miners inebriated.

11 Van Onselen, *New Babylon, New Nineveh*, p. 106.

Was it for this reason, that, in that same year,1906, the East Rand Mines Proprietary Company paid out wages for the brewing of 'kaffir' beer to a woman named Bella, the only woman, other than Ellen Richards (who found a cheque), who is accounted in the Cash Book? And was it for this reason that the ERPM ordered a bag of 'kaffir corn', from one R. I. Innes in August (p. 57v)? Who was Bella? She appears under the heading 'Native Compound'. Where did she come from? Did she run a shebeen near the miners' quarters, hold sway over the men who came and went at the end of the working week? Did she give them credit, keep their debts on file in a cash book of her own? When she was paid, did she walk to the mine management office to receive her cash—£2.10 in December (p. 84r)—and hide her money in the folds of her skirt on the way back, in that time of Jan Note?

Once again, the Cash Book leaves us to speculate. And, speculating, we can note that Bella can mean either beauty (in English) or conflict (in Latin); the name folds over onto itself in yet another uncanny echo of gold in the chamber of mines.

Native Advances, Active Service

In 1906, the Union of South Africa had not yet been consolidated, and though the conclusion of the Anglo-Boer War had ensured British authority in the Transvaal, the integration of the British colonies and the Boer Republics was still in process. The year of the Cash Book was in many ways a crucial transitional moment. For, when, in 1906, the lords of the Witwatersrand leveraged new laws to enforce the addition of a poll tax to the hut tax in Natal, hoping to force the last rural men to abandon their diminishing farms for the mines and in this way compensate for the loss of the departing Chinese labour, Bambatha kaMancinza (*c*.1860–1906) led a revolt against the authorities. The Bambatha uprising was in some ways the first real test of the new confraternity of colonial state and mining capital, which, prior to the war, had to compete with the powerful lobbies of Afrikaner agriculturalists. But when it was over, the pathos of the victory for those named Native would be narrated by the Zulu poet B. W. Vilakazi in the language of a deformed pastoralism: 'We agreed to leave our kraals: / In herds we came—castrated cattle!'[12]

12 B. W. Vilakazi, 'Ezinkomponi', translated from the Zulu as 'In the Gold Mines' by A. C. Jordan. It originally appeared in the collection *Amal' ezulu*, Bantu Treasury Series VIII (1945).

The new state—forged in fire, one might say—cooks its books and leaves us its mark in the ashes, with the calculation of wages paid to the miners who have abandoned the shafts to fight for the British against the 'Zulu Rebellion'. They are carefully noted in the Cash Book. December 1906: 'Paid sundry employees one month's wages for going on Active Service in Zulu Rebellio'. The cost: £734. It is the largest single payment for wages noted that year.

There is no mention of the African miners who might have fled their contracts to support the uprising, though one surmises that some at least dreamt of joining the insurrection. Perhaps they are ghosted in the entry that describes the 'Natives at D & A who have deserted' (p. 58v). Certainly, the Native workers had abandoned their employ for insurrectionary war in the past. When miners had been paid in guns in Kimberley, before the standardization of wage labour, their flight from the diamond mines to the fields of blood had been massive indeed. Exactly three decades before the Cash Book was opened, the Pedi workers in Kimberley had taken their 'hard earnings' and fought against President Burgers of the Suid Afrikanse Republik, leaving the diamond pits deprived of nearly half its workforce. By 1906, however, such a mass exodus was impossible. Perhaps even unthinkable for all but those who could join Jan Note and his mobile armies.

Nonetheless, there are others whose absence is both noted and remunerated. In the space just beneath the reference to the sundry workers paid for their service in the fight against the Zulu Rebellion, there appears the name of Mungul Singh. Who was Mungul Singh? Was he himself a long-term resident of Natal, one of the many Indians who had come to South Africa during the era of indentured labour, not as a labourer (they were mostly from Gujarat, and not often named Singh) but as a member of the British-educated subaltern class that would later prove so useful to the architects of racial division? Or had he come more recently, with the other Sikhs who served in the

British armed forced during the Anglo-Boer War? Still other Indians had come as merchants, heirs to the long tradition of trade that extended the centuries of exchange between the Indian Ocean and the interior lands. Many had sought opportunities in the textile trade or operated laundries that competed with the Zulu men's washing guild, the *amaWasha*, which, between 1890 and 1906, constituted one of the most successful forms of black corporate activity in the Witwatersrand.[13]

13 Van Onselen, *New Babylon, New Nineveh*, pp. 275–308.

From Natal originally, the *amaWasha* actually attired themselves in an Indian fashion. They were famous for their turbans, as well as their exemption from the Masters and Servants' law. Van Onselen has described their entry and forced exit from the life of the male-dominated city of Johannesburg in some detail, noting how, in 1906, a series of factors conspired to propel many of the *amaWasha* back to Natal. To begin with, segregationist legislation forced Zulu men out of the city (mainly to Klipspruit, on the West Rand, near Soweto), where business was hard to come by and water supplies were contaminated by the effluent from the mines. But the entry of cheaper Chinese laundries into the field of competition also repressed income and made staying on the Witwatersrand difficult to support. Finally, the news of the Bambatha uprising added fuel to the exilic fire, inducing either enthusiasm for revolt or fear for personal property in the care of female householders.

Did it seem odd to Munghul Singh to find the forms of his own still-colonized nation imitated by the Zulu washermen, so far from India? Did it make the fight against the Zulus more or less difficult when they appeared not in turbans but with *asegais*? Whatever his origin, and whatever his thoughts, Mungul Singh was paid along with the other anti-insurrectionists to fight these Zulu men. It is a painful testimony to the absence of something like an 'Afro-Asian people's solidarity', which would, in any case, only finally be realized—and ever so briefly—at Bandung, nearly a half century later. For his part, Singh may have been encouraged by Mohandas K. Gandhi, whose Natal-based newspaper *Indian Onion* was urging Indians to fight against their black co-residents in order to receive full recognition as citizens of the empire. It is sobering, and somewhat demystifying to realize that Gandhi was urging these Indian men to follow Achilles to stain the dust with blood in the very year that he invented the concept of satyagraha to denote the tactic of non-violence.

Insufficiently flattered, the British refused their Indian subjects' service except as stretcher bearers on the fields of death. So Munghul Singh may have been one such bearer of wounded bodies, unless as a soldier already in Her (and then His) Majesty's service in the Anglo-Boer War, he was permitted to actually do battle. It is possible, of course, that he was merely an adventure seeker or someone who found the life in the mining compounds stifling or someone who wished to return to the coastal environs where the winters did not require so much coal and the summers were less arid. We know only that he was there, on the side of the British, and that he returned to receive his due and the imperial gratitude of the ERPM.

Stamp Required to Replace Same

Unmarked in the columns of the Cash Book, but etched there in the wages of the volunteer soldiers, the official tally of the Bambatha uprising rose to at least 3,500 Zulu men dead in that year, 1906. Twenty-four white soldiers also lost their lives. The head of the defeated Bambatha was paraded as a trophy, though rumours of his survival and escape continue to circulate even today. In 1906, a photograph of a severed head named Bambatha was transformed into a postcard and circulated internationally. Exactly one century later, in 2006, the image of that decapitated visage returned, this time as an uncanny image of the newly titled *hero*, Bambatha, to adorn not a card but a postage stamp, the original medium and currency of the national geo-body.

After 1906, and the death of Bambatha, the collection of taxes proceeded without significant resistance. Evicted and alienated, and comprising an increasing percentage of mine labour, as the Chinese workers began their sojourn home, the African Natives would nonetheless be refigured as people of the land. In forced cyclic migrations that linked country and city via the phenomenon of the peasant-proletariat, they were stranded in the demand for eternal returns. Until death. Or until someone else could replace them. The organizing of mine labour via restricted movement, short-term contracts, and the cultural enclosures that would later be called 'homelands' or 'bantustans' initially secured for African labourers a certain freedom. If they found the conditions on the mines unbearable, if they were able to satisfy their cash needs or, in the early days, if their payment in arms and munitions had met the demands of their chiefs, they could return home, and although breach of contract could lead to imprisonment, it was generally possible to refuse mine work. Mine management complained bitterly about this capacity of southern African miners to 'desert' but had few means to combat it—except to import others. They could nonetheless be grateful that such desertion tended to obviate the other possibility, namely collective action aimed at the improvement of labour conditions and better wages. The white miners were far more likely to strike. So, as it turns out, were the Chinese labourers, those 'Coolies' who appear so repeatedly in the Cash Book, and whose substitution for Natives finds its perfect graphical allegory in Cash Book's crossing out of the word 'Native', followed by the corrective word 'Coolies'.

~~Native Labour~~ (*Coolies*)

For a brief period, but mainly between 1904 and 1906, the gold mines of South Africa undertook an enormous experiment in global labour importation, following the Californian example, bringing more than 63,000 Chinese workers to the Witwatersrand. The final departure of these indentured miners occurred in 1910, but the last group of recruits disembarked at Natal from the *Tweeddale* in November 1906—a total of 2,129 workers, including three women and four children. What terrors may have afflicted those three women on a ship of more than 2,000 men we can only surmise. Indeed, it is unusual to discover in this and other ships' manifests, the names of women. Nonetheless, ERPM embraced them. And, from Cason mine, an application was sent in August to establish quarters for Chinese families. Clearly, the women on the *Tweeddale* were few but not alone.

The last arrivals had not been anticipated when the licence for Chinese family quarters had been made and many thousands more workers were yet to arrive on Natal's shore from one of the recruiting centres across the sea. Four months later, in the Cash Book monthly accounts for December 1906, the first 9 of 10 entries concern Chinese labour, including capitation fees, passage fees, repatriation fees and general expenses.

The campaign to import 'Asiatic' labourers was led mainly by ERPM's chairman George Farrar. In fact, ERPM employed more Chinese labourers than any other company on the Witwatersrand, accounting for almost one half of all those brought to South Africa. It is for this reason that the Cash Book, whose very architecture anticipates their presence, is so redolent of references to these migrants.

The supply was coordinated by the Chamber of Mines Labour Importation Agency, established for this purpose in 1904 to take over from the Committee of Agents set up in the previous year, with capital that was pooled from all of the main mining houses on the Rand. Private companies competing for the transport contracts sometimes referred to the Chinese workers as 'Freight on Board', barely distinguishing between the men and the supplies that were said to be necessary to their well being: padded coats, Chinese fireworks, gongs, joss sticks, chopsticks and rice bowls, conical hats, preserved ginger, dried fish, special Chinese writing paper, Chinese wine, paper fans, special cooking ovens, bamboo shoots and even pipes and opium. A. P. Cartwright recounted this archaeological dream image of Chinese everyday life, but if these special considerations were much discussed by the proponents of Chinese labour, they find no echo in the Cash Book.[14] It seems that rice was rarely provided—there is instead a notice in April of a telegram sent to Mansel and Davies regarding the unsatisfactory state of the meals being served in the Chinese compound (p. 21r). And the traffic in opium was prohibited—though at least one compound manager was dismissed for selling it to the Coolies in his charge.[15] The only reference to clothing appears in the 'boots' and 'clothing' that are noted in February (p. 2v) under the heading 'Chinese Advances'. In other words, the Chinese workers were charged for them, the costs of surviving the Highveld winter added to the other costs, such as transportation, food and coal for heating, which had to be repaid prior to repatriation.

What mad dream led the mine magnates of the Witwatersrand to seek their workers on the other side of the world? It is as though the childhood game of digging all the way through the earth to China was being literalized by the South African mining companies. The ostensible rationale was a shortage of

14 A. P. Cartwright, *Doctors of the Mines* (Wynberg: Purnell and Sons, 1972), p. 10.

15 Jeeves, 'The Control of Migratory Labour', p. 16.

African labour. Indeed, the number of African workers in the mines declined by about two-thirds after the onset of war in 1899, as did gold output. Following the British victory in 1902, the need to rebuild the local economy following the war depended on getting the mines up and running as quickly and as cheaply as possible. But a host of factors were making mining more difficult and more expensive, from the dissipation of surface deposits to the fall in gold prices on the international commodities market and the increase in transportation costs. Investors in London and New York had begun to pull their money out of the industry, drying up capital reserves and squeezing South African companies even more.

No doubt, the black South African workers could have been forced to return to mining, or indeed, to enter the field anew, as they were following the Bambatha uprising, but the political energies of the new government had other local priorities. And from England, Milner's government was pleased to see a solution that was colonial in scope, one that would help knit together the far-flung territories of the already disintegrating empire. Farrar's case for the Chinese workers depended on their relative affordability (the wage rate in China being so low as to make those paid in South Africa seem attractive). It was also motivated by his concern over the very high rates of death due to lung disease among African workers from further north on the continent, a fact he attributed to the constitutional weakness of 'tropicals'. But he did not advocate the permanent settlement of these migrant workers. To the contrary, and in a manner that resembled the policy on Native labour, the importation of Chinese workers was to work by cycling men in and out of the mines, keeping them moving, albeit on longer-term bases, between home and abroad.

The system for importation was elaborate and, as it turns out, expensive—far more expensive than that designed to recruit African labour. Mining companies calculated that their outlay was about £17.5.2 per Chinese mineworker, not including initial capital outlays (including the mandatory purchase of shares in the Labour Importing Agency). Once advance wages and capital outlays were added, the cost went up to £31.10 per person. It was at least £7 pounds more per head than the costs associated with African labour, and for this reason, the mines were concerned to maintain the state of productivity of their workforce—by hook and by crook, as the saying goes.[16]

16 Jeeves, 'The Control of Migratory Labour', p. 16.

To that end, the mines were prepared to import Chinese interpreters and doctors, even operating a medical school for the latter in Tongshan. The goal was a ratio of one physician for every 200–300 workers, paid at a rate of between £4 and £8 per month.[17] Doctors or not, more than 1,100 Chinese miners died on the Witwatersrand during those few years. Their deaths are calculated in the unpaid wages that accumulate in the columns of the Cash Book and in the purses found on corpses, described in the asides of the blank column. Early on in the discussions about importation, the mines had contemplated the possibility—and the costs—of embalming the Chinese miners and returning their bodies home, but this idea was never realized; instead, there are burial expenses accounted in the Cash Book. Repatriation was apparently available only to the living. Nonetheless, Chinese conceptions of death preoccupied the magnates. It was said that the Chinese workers feared dissection after death and that they believed the white doctors would use their eyes for medicinal purposes. For that reason, doctors were advised not to close the eyes of those who died in the mines. Instead, they were left to stare eternally at the sky from which the Holmes Comet had vanished, and to go unblinking to their gods.

17 Cartwright, *Doctors of the Mines*, p. 12.

According to the *Rand Daily Mail*'s reporting in March 1906, Chinese workers died at a rate of about 18.5 per 1,000 each year, compared to the 47 per 1,000 that was reported for Native labourers. The disparity was one of the reasons for the belief, belied by so many other indicators, that the Chinese labourers were treated much better than were their African counterparts, who were felled quickly by pneumonia and other lung diseases following their less rigorously supervised transportation along the East Coast of Africa in ships that had often enough first plied those waters with slaves in their holds. In the barracks built for the Chinese workers, short cement beds resembling shelves or perhaps mortuary slabs were stacked one atop the other. More than one observer, seeing these slabs today, has assumed that they were the model for the bunks of the concentration camps—and, in a way, as Van Onselen makes clear, they were.

Was this shelving of labour one of the reasons that the Chinese workers in the Cash Book appear so often as the Coolies and Chinese defaulters escorted from jail after refusing to work in the miserable depths? Several times, the Cash Book records expenses for the transportation of deserters, caught and imprisoned in the very barracks that would supply convict labour on other occasions. It redoubles the reportage of the newspapers that often recorded the disappearance of the Chinese workers in the Highveld night. A particularly florid tale in the March 10 edition of the *Rand Daily Mail* narrates the 'Wandering Chinese', whose flight from the North Randfontein mine led to mounted patrols and 'telephonic messages' being sent after them. It concludes, 'The whole of the district was scoured but none of the Chinese were met with.'

In order to escape a contract legitimately, a Chinese worker had to purchase his discharge. His repatriation fee—which was £17.10—was to be paid out of his own savings from the time of his labour in the mines. If a worker could accumulate such funds before his contract was out, he could purchase his liberty. In the Cash Book, these compromised efforts to buy back sovereignty provide the rare occasion for recording the name of a Chinese worker and not merely his registration number. The Cash Book recognizes, for example, Coolie Jong te Jim, No. 40041, who purchased his discharge at a rate of £23.15.10, and Coolie Reghs, No. 2420, who paid £18 for his. These were vast amounts of money for a migrant worker. One imagines the cash stored in little bags around the neck, tucked under pillows and mattresses, hidden in boots and under teapots, buried. Another miner might have loaned it to others, gambled it away, doubled it at the mahjong table, spent it on trinkets or meals at the local tavern or in the brothels adjacent the compound, or used it to invest in a laundry. Jong te Jim and Reghs had instead saved it, hidden it from bandits and bribe-seeking bosses and used it to purchase a hastened return to China. They were lucky.

On occasion, a miner died before he had the opportunity to pay for his freedom. Such was the case with Wang Hai, No. 55917. When £3.2.6 was found in a purse on his person, the ERPM remitted a mere 5 shillings, 11 pence as the 'balance' owing to the Chamber of Mines (p. 92v). It was what remained after the repayment for those boots and clothing noted earlier, items which could only adorn the man's corpse with the uselessness of ritual.

But not all the miners desired to return. Sometimes, too, there is a reference to a Chinese worker who chooses not to go home at the end of the contract, whether for the transgression that was love in South Africa or for fear of even worse exploitation at home or because he had discovered new economic opportunity on the Highveld. Such was the case with Coolie No. 48791, who enters the Cash Book after requesting a refund for the 'amount paid by him for purchase of discharge [as] he does not now wish to avail himself of the privilege of returning home' (p. 86v). It is an astonishing entry, this single statement, written in a prose that attests to the significance of the event but suffused with a tone of supercilious surprise at the Chinese man's temerity. Who was Coolie No. 48791? Where did he go? Into what shadow of the South African immigrant world did he vanish? Perhaps he stayed with Coolie No. 48834, who is entered on the previous page, having been similarly refunded for the purchase of his discharge, 'he being desirous of remaining' (p. 84r). Their numbers are so proximate that it leads one to believe that they were recruited together, and that they travelled to South Africa on the same ship, disembarked at the same point and entered the same mine upon arrival—to endure hardships and successes, loneliness and discovery together. Friends? Brothers? Cousins? Lovers?

An Undesirable Boss (Coolie)

In August 1906, ERPM paid a Native employee to travel to Johannesburg with an application to repatriate an 'undesirable Boss Coolie'. It is an unusual reference in the Cash Book, most of the entries concerned with discipline having to do with 'wandering' or deserting workers and their transport back to the mines from jail. Was the Boss Coolie one of the ringleaders of the increasing number of disputatious gatherings that were occurring on the mines of the Witwatersrand?[18] Or was he one of the brutal enforcers who had been brought from China to serve as both policeman and translator, and who had grown delirious or cruel with the power vested in him? Both figures haunt the spaces between the lines about the defaulters and the deserters, the gamblers and the dead men.

Between 1904 and 1907, there were several significant work stoppages organized by Chinese miners, almost all of

18 Gary Kynoch, 'Controlling the Coolies: Chinese Mineworkers and the Struggle for Labor in South Africa, 1904–1910', *International Journal of African Historical Studies* 36(2) (2003): 309–29.

them concerned with wage disputes, the transformation of incentives into penalties, and the imposition of new minimum productivity requirements (such as a minimum number of inches to be drilled per shift). Sometimes, the work stoppages led to the riotous destruction of white miners' accommodations, as perhaps occurred in April when the Cash Book notes a payment to 'G.J.V.D Merwe for damage done to his house by Chinese' (p. 26r).

Rations were sometimes withheld after such protests or if workers failed to meet their daily minimums and it was in response to such debilitating punishments that the Chinese workers often took action. Not incidentally, it was at ERPM's own Cason mines that the last strike by Chinese workers would be undertaken in 1907, only a few months after the last migrant had disembarked the last ship form one of the treaty port cities of China.

The policing and management of 'undesirable Chinese' fell to the Chinese police, but it was overseen by the Foreign Labour Department, a nominally autonomous institution that had the capacity to indict, prosecute and penalize workers for infractions of pass laws, failure to show for work, gambling or any number of acts deemed offences under the ordinances of the time. Between June of 1905 and June of 1906, when notations appear in the Cash Book on a nearly daily basis under the heading 'Chinese Defaulters', a total of 11,754 offences by Chinese miners were tallied, including 1,700 cases of desertion. Sometimes, of course, desertion simply meant getting lost on the way back from town, when during the one day off a week (a workday lasted 10 hours), many of the miners took consolation in the brothels and the bars beyond the compound's perimeter. Magistrates were dispatched to mine property every few days to try these cases, which were so numerous that the regular courts would have been drowned by the caseload. It is for this reason that, in August of 1906, ERMP paid the postage for a license to build a court exclusively for the trials of Chinese offenders (p. 42v). What justice might have been meted out can only be pondered in the idiom of parody.

Until 1905, flogging had been a common method of punishment for those found guilty (the vast majority of those accused). But it belied the effort to distinguish between slavery and indenture, and partly for this reason, the predicament of the Chinese Coolies had been a scandal in the British Parliament from the earliest days of its conception, when Lord Milner and Winston Churchill had debated not only the costs and responsibilities of the project but also the anachronism and the hypocrisy of this new form of servitude. News of floggings competed with claims to special treatment, the two contradictory images forming something like a stereoscopic doublet that refused to be resolved into a single vision. English liberalism was, at the time, a politics of astigmatism. Nonetheless, after Lord Selborne replaced Milner in 1905, 'Imperial Fluids' were granted to cover the fares for those who had not amassed the modest sums required for return. The experiment in global labour substitution would come to an end in the year of the Cash Book, appearing as either a stopgap or an error in the history of the mines of South Africa.

Correction in Red

In the Cash Book, mistakes noted by the auditor are overwritten with red pencil. In history, mistakes may be acknowledged but not eliminated. Every attempt at erasure leaves its mark, albeit one that is difficult to interpret. The termination of the Chinese labour experiment had been aided and abetted by an improbable consortium of English liberals, Afrikaner nationalists and Jewish emigré merchants. In the year that the Dreyfus affair finally ended in France, Jews appear in the Cash Book only in the shadow of a name—G. W. Solomon, who is regularly paid for transporting gold to Johannesburg.

What appealed to one group repelled the other, namely the depressed wages of the Chinese workers. If low remuneration meant relative profit for the mining companies, it meant little revenue for the merchants. They complained vociferously about the lack of patronage for their goods—whether foodstuffs or entertainments—though they were as likely to blame the stringent consumption of Chinese workers on cultural factors and a putatively natural modesty of appetite, as on depressed wages.

Such concerns had been expressed long before the systematic importation of Chinese mineworkers had commenced. They had been enumerated, for example, by Meyer Dovid

Hersch, a Lithuanian Jewish immigrant who had come to South Africa a few years after the discovery of gold, and written a series of travelogues and popular histories about the people of South Africa for publication in the Warsaw-based Hebrew-language newspaper *Hazefirah*. These ethnographic miniatures had derided the Chinese for being 'content with just one outfit of white clothing and a little rice from time to time', which fact 'brought onto them the wrath of the other business people'. Hersch's ventriloquism of Jewish resentment had become even sharper when it appeared in a 1903 series of 'Letters from Transvaal', wherein he wrote of the tension between the Chamber of Commerce and the Chamber of Mines: 'The Chinese labourers who were brought into the country are men of extremes who eat frugally and whose expenses are very small; sufficient for their needs is a handful of rice which lasts from one week to the next. It is for this reason that they are prepared to work for so low a wage'.[19] Hersch was therefore appalled when the Chamber of Mines determined to bring in even more Chinese workers, and chastised the mine owners for not considering the consequences of their strategy on local business, concluding that the result would be that 'the wealthy will indeed grow richer and more powerful' but that the 'middle class'—'a large number of whom are Jews'—would suffer.

Ten years before the Cash Book was opened, in the year that Hersch had written an account of 'The Rebellion in the Transvaal and its Relation to the Jews', about 10 per cent of Johannesburg's population was Jewish, many emanating from the northern Lithuania district of Kovno that Hersch had once called home. He must have been relieved, in 1906, to learn that there would be no more Chinese coming to the mines of the Witwatersrand. But he had been wrong to imagine that the Jewish men among the magnates would bear their religious confreres in mind when conceiving of labour policy—as he did when he concluded his Letter No. 2 by putting his hopes in 'the three biggest mining magnates, Barnato, Eckstein and Langermann,' whom he assumed would 'employ black workers, at a respectable wage'.[20] As it happens, Eckstein was the son of a Lutheran minister, and not a co-religionist at all. But on the Rand, as elsewhere, class interests trumped those of religious solidarity in any case.

19 Joshua I. Levy (ed.), *The Writings of Meyer Dovid Hersch: Rand Pioneer, Historian and Recorder of Jewish Life in Early Johannesburg* (Johannesburg: Ammatt Press, 2005), p. 118.

20 Ibid., p. 143.

George Farrar's ERPM Board nonetheless had a number of prominent Jewish members. Among them, Beit, who was on its London Committee, was the son of well-to-do Hamburg traders, who, having gone initially to Kimberley, became a protegé of Cecil Rhodes. Lionel Phillips, the child of Jewish merchants in London, arrived in Kimberley at the same time as Beit, and was similarly enamoured of Rhodes's ambitions. With the Albu brothers, Barney Barnato, Joel Solomon, Gustav Imroth and others, they were members of a remarkable transnational transformation that would see many of them move from the status of German Jewish emigré to British baron, often enough via the detour of treason against the Transvaal government in the Jameson conspiracy.

The Jewish Randlords who had grown up in trading families no doubt enjoyed a familiarity with mercantile life that prepared them well for the speculative frenzy of the diamond and gold mining worlds, which initially depended as much on buying and selling claims as on the more complex financing of productive mining. But, in truth, most of the Jewish immigrants to southern Africa were, like those described by Hersch, small-scale traders whose livelihoods depended not on the management of large capital but upon the consumption of workers employed by the mines. Moreover, the division between the upper and lower classes of the Jewish population in the Transvaal tended to mirror and reproduce a national fissure. It was split between Jews of Russian or Lithuanian descent, many of whom were petty traders of orthodox and Zionist tendencies, and the English and German Jews whose representatives were often to be found in the less orthodox Anglo-Jewish Federation and who were more prominent in the upper classes of the mining aristocracy. Among the so-called Randlords, only Samuel Marks was of Lithuanian origin, his initial flight to England occasioned by fear of persecution in the waves of anti-Jewish violence that arose in the aftermath of the assassination of Tsar Alexander II in 1881.

News of the massacres of Jews in Russia appeared not infrequently in the *Rand Daily Mail*, the newspaper read by all of the mining world. In March of 1906, it was included in a report about a motion in the British Parliament condemning such violence—in a proclamation of British concern for the recognition of the rights of all peoples. It sits awkwardly to an adjacent column on the same page in which the British suppression of Zulus at Mapumulo is described along with the growing concern about the disquiet of Chief Bambatha—that suppression to which Gandhi had summoned his Indian compatriots in the ironic hopes that they could finally earn the recognition that would grant them citizenship in the empire.

Recovery of Coolie

Following the departure of the Chinese miners, the word Coolie would revert to its use as a descriptor of South Asian workers. But even in the Cash Book, it encompasses these other 'Asians' from a continent not yet imagined as one. There is Rangaan, who is paid wages in February and June under the Engineering Department's account (p. 42r). And Coolie Ramsammy, whose debt of 7 shillings, 6 pence is identical to that of Lerche, Trounson and Roff, names which follow. This was the amount owed for coal, though Ramsammy's account is described in terms of the store to which he owed his money, not the commodity for which it was spent (p. 19v). Where Rangaan worked and in what capacity is a mystery. Nor can we surmise from the Cash Book whether he was a relatively skilled worker or someone who, like Campbell, worked in the garden. His wages seem too high to be those of a mere sweeper or sanitation worker, but they are also hard to isolate, as they appear combined with those of Campbell, totalling £4 for the month.

No doubt, Campbell and Ramsammy, named and exceptionalized in the world of Natives and Coolies, received occasional bonuses, as at Christmas when, in the Cash Book, costs are calculated for Christmas boxes to the 'office boys' and 'Natives at office'. Even so, the only overtime noted in the entire volume, other than that for the prison wardens who oversaw the convicts at Cason, is that paid to J. Llewelyn, for working on both Christmas Eve and Day. Mine shifts were 10 hours per day, and the working week was 6 days long in 1906. For ordinary mineworkers, there were no holidays.

Amounts Collected

Over time, the scale of ERPM's transactions accumulate. On a daily basis, however, the company's activities can seem slight in the Cash Book, where the tallying of costs for a mule harness is as meticulously inscribed as those for the payment of insurance upon the death of workers. The scale of operations is disclosed in the crowds of workers whose numbers are totalled en masse in the Cash Book. Recurring across the months like a stutter is the reference to passes, those documents whose possession either enabled or inhibited movement, often substituting for every other biographical detail by which a person might otherwise have been known. The enormity of the ERPM's undertakings is intimated on the first page of the Cash Book, where costs are tallied for 5,725 passes for Native workers. Later, there are entries for 60, 17, 13 and 7 more passes. A trickle of men after the deluge. But by April, the second month to appear in the Cash Book, there is need for more workers and another 5,624 passes are purchased at a cost of £564. Each month, thousands of men made their way to Angelo, Cason, Comet and Driefontein, to Cinderella, Hercules and Angelo mines, to be processed and assigned a task appropriate to physical capacity. On trains and by foot, they moved in nearly constant trails of human bodies, to be counted and numbered, numbered and accounted. When the Chinese workers arrived, they, too, came in huge numbers: 1,943 arrived on the *Indravelli* and 1,590 on the *Cranley* alone. Men not like mountains but like heaps of bodies, moving between the upper world and the underground, but always bound to the earth.

To orchestrate these mass migrations was no small feat, and, imperial support or not, the Chamber of Mines Labour Importing Agency (CMLIA) was not capable of coordinating the complex recruitment, supply and transportation of workers from the diverse regions of China's vast interior. It was already taxed in the Portuguese territories. And so, it contracted with Western companies who were operating out of the treaty port cities, and they, in turn, contracted with Chinese agencies and recruiters, using a combination of diplomatic authority, the lure of commissions and bribes, to achieve their goals in a climate that was not very amenable to emigration.

In China, the CMLIA was based in Tientsin City, a British concession that was also serviced by a branch off the main Tientsin-Manchurian Railway that gave the company access to Ch'inghuangtao and Chefoo, and to cities north. It was the perfect netting point for a broad array of labour-migration routes. It was also the seat of the Chinese Engineering and Mining Company, which not only managed the railway but also served as a major recruiter for the northern provinces, a role it shared with three other foreign firms—Cornabe Eckford & Co., Silas-Schwabé & Co., and William Forbes & Co. The CMLIA canvassed the central provinces via another set of companies, namely Jardine Matheson & Co., and Gibb, Livingstone & Co., while Butterfield and Swire recruited in Hong Kong and Swatow.[21]

21 Richardson, 'The Recruiting of Chinese Indentured Labour', pp. 95–7.

These companies had supplied labour to other transnational projects—including the trans-Siberian railway—and were engaged in a wide array of activities, such as the trade in hides and wool, but it was the Chinese Engineering and Mining Company's engagement in activities like those of the Chamber of Mines that gave it the edge, and made it so significant in the story of the East Rand. A mining company, one presumes, knows what a mining company wants and needs.

The general manager of the Chinese Engineering and Mining Corporation, which supplied ERPM with so much of its Chinese labour, was an American geologist and mining engineer named Herbert Hoover, the one-time translator (with his wife, Lou Henry) of the great sixteenth-century treatise on mining and mine management by Georgius Agricola, entitled *De Re Metallica*. Georgius Agricola was the pen name of George Bauer, but the pseudonym also functioned as a translation. Bauer means peasant or farmer in German and George kept the reference to agriculture in his new Latinized surname, even as he turned his mind to more subterranean matters with a lucidity and insight not to be surpassed for another 200 years. What stands out, even today, is his insistence that mining and agriculture are linked, not only via the migrant labourers who vacillate between ploughing the field and plumbing the earth but also in their mutual dependence and coeval origins. 'Without doubt, none of the arts is older than agriculture, but that of the metals is not less ancient; in fact they are at least equal and coeval.'[22]

22 Georgius Agricola, *De Re Metallica* (Herbert Clark Hoover and Lou Henry Hoover trans) in *The Mining Magazine* (London: Salisbury House, 1912 [1556]), p. xxv.

If mining and agriculture were ancient, however, finance was not. The new arts included the creation of joint stock companies to distribute the costs and risks of industrial mining, which Agricola advised even in 1556, exactly 450 years before the Cash Book was opened—a year still noted for the catastrophic earthquakes in Shanxi Province, China (and the consolidation of the Mughal Empire under Akbar in India). One can imagine that Hoover's fascination with Bauer's text, which he admired for its poetic language as much as its geological insight, was provoked in part by Agricola's concern with risk management. And if Farrar had read Bauer's text in Hoover's translation, he might have thought, if only for a moment, that the ERPM had been the brainchild of the early modern philosopher of mining, for it was a pure product of the financialization of mining that had emerged in Kimberley and then perfected itself on the Rand.

In Agricola's time, as in our own, mining was both a risky undertaking and a morally questionable one, and the fear of losing everything was assuaged, at least partly, by the fantasy of nearly infinite wealth. It was for this reason that mining—and especially gold mining—had seemed, to many of Agricola's contemporaries, to be not merely the art of going underground but, rather, of descending into hell itself, to the place where evil tempts men to avarice and war.

A learned man of unfailing Catholic commitments even in the time of Martin Luther's Reformation, Agricola turned to Virgil's *Aeneid* to answer the critics who blamed the mines and their products for the violence of wars and the rapacity of thieves. He invokes Virgil's description of Polymnestor's murder of Polydorus for an image of what a love of gold can make a man do: 'To what wilt thou not drive mortal hearts, thou accursed hunger for gold?'[23] And Agricola answers, at once summarizing Virgil's complaint and offering a tempered defence of mines against men:

> [L]ust and eagerness after gold and other things make men blind. [. . .] Moreover, those who have been so addicted to avarice as to be its slaves have always been regarded as mean and sordid. Similarly, too, if by means of gold and silver and gems men can overcome the chastity of women, corrupt the honour of many people, bribe the course of justice and commit innumerable wickednesses, it is not the metals which are to be blamed, but the evil passions of men which become inflamed and ignited; or it is due to the blind and impious desires of their minds.[24]

23 Ibid., p. 16.

24 Ibid.

When the liberal opponents of Farrar's Chinese-labour strategy raised their objections in the British Parliament, it was not servitude to gold that they worried over but the enslavement of those who produced it. They owed their capacity to do so at least partly to Agricola and his treaty on mines and men, and partly on Hoover's translation.

Hoover, who had moved to China after working for Bewick Moreing in Australia, would become US president in 1929, as the world was about to be plunged into the first truly global financial crisis, an event that many have linked to the reestablishment of the gold standard after its abandonment during the Frist World War. He is not named in the Cash Book, but can we say he is not there, exercising as he does this force from afar?

Accident Assurance

Hoover was nominated for the Nobel Peace Prize many times, mainly for his work supplying food to war refugees following the Frist World War. But Nobel appears in the Cash Book as the supplier of dynamite. The explosions underground, so necessary for the liberation of rock from its earthen home, were the cause of many accidents, of course. Roofs collapsed, rock walls burst open, underground rivers swept the miners to subterranean seas. Limbs broken, hands crushed, noses severed—the most common of injuries were defacements. What was a torture elsewhere was, here, a mere accident—if, that is, one can use the term accident underground.

Nineteen hundred and six was a year of accidents. Terrible accidents, of terrifying scale. At the Courrières Coal Mine in northern France, a dust explosion caused either by the accidental ignition of explosives or by methane gas lit by an open flame killed 1,099 miners. It was the largest mining accident in recorded history, and not only was it reported in the press to an unprecedented degree but photographs of the disaster were also quickly mass-produced and, like those of Bambatha's head, turned into postcards, which were then sold to raise funds for victims' relief. The accident, which occurred on 10 March, was followed a week later by a regional strike, which nonetheless failed to elicit substantial changes in the French coal-mining industry.

In the *Rand Daily Mail*, that register of South Africa's mining world, the event was reported within two days with uncharacteristic emotion as 'One of the most terrible disasters ever recorded in the history of mining . . .':

> A fearful explosion of fired-damp occurred, destroying the cage of the principal pit, and entombing 1,050 men, who were at work in different parts of the mine.
>
> The explosion brought thousands of men, women, and children to the pit's mouth to learn the extent of the disaster, and some heartrending scenes were witnessed.
>
> Rescue parties immediately organised, but for some time, the brave fellows were unable to approach the pit's mouth owing to the intense heat—the main shaft resembling a huge furnace—and the suffocating 'choke-damp' arising from the explosion.[25]

25 *Rand Daily Mail*, 12 March 1906.

There are no such narratives in the Cash Book. The closest it comes is the recognition of the monies paid as reward to men who 'attended the fire' at the South Shaft of Angelo Mine on 31 May, less than three months after the colliery disaster in France. It is left to us to wonder if there were, on occasion, similarly heart-rending scenes. But it is clear that there was no community of women and children who could come to the mine's edge to call out after the vanished. The Native and the Chinese workers were almost all severed from the intimate worlds that might have formed this mournful cordon.

The Chamber of Mines has no official figures for accidents before 1911, but they are here, in the Cash Book. On the recto of the opening page, under Accident Assurance Account, payments are noted for the 'full settlement' of the claims made by European miners named Stone, Wickham, Heltne and Stocking, the latter receiving the remarkably large sum of £77.13.5. Not everyone lived to make claims, of course. In August, for example, the deaths of Natives at Driefontein and Comet mine are factored in at a cost of £1.7.8 and £10.6 respectively. Month after month, the Cash Book records the payouts for the wounded and the compensations offered to the families of the deceased, measuring the value of life in pounds, before seeking the final balance. And even so, the sum is missing.

The new system of keeping accounts that E. H. Blackmore and O. Meredith learnt in Pretoria, their travel paid for on 1 October (p. 65r), was not one concerned with the numbers of wounded but with the monetary records of workshops. But later record-keeping gives us a sense of the scale and frequency of the accidents for which insurance provided its belated compensations. Between 1911 and 1994, 69,000 miners, mainly black, were killed in South African mines and more than a million injured. And that doesn't include the slow, unfurling wounds of lung disease, which would generate mountains of corpses over the century despite the fact that, already in 1906, Albert Calmette and Jean-Marie Camille Guérin had developed the first vaccine against tuberculosis. When one tallies the dead that accumulated without immunity, the numbers are more typical of a war. As though commanded by Shylock, pounds of flesh were literally materialized in the gold mines, if only to be converted to sterling.

Received from Time Keeper

Holding sway even above Hercules, the actuary is the petty god of this sordid world, where the calculation of risks and benefits shrouds the beneficiary—capital—in a mystic cloud. Against the actuary, the sangoma's sorcery and the rituals of muti offer their palliations (every mine has its shrine and the obeisance to ancestral spirits underground is a feature of many miners' lives), but they do not keep the ghosts at bay. For accident is the horizon and the spectre that is always awaited in the mines. This waiting cannot prepare itself for the eventuality that will terminate it. Rather, it constitutes a second order of time, a duration without futurity, beyond the cyclicity of migrant labour and without the punctuation of the Cash Book's calendar, with its days and months in measured sequence.

Peter Abrahams captures something of the quality of that enduring-not-waiting, what we might call the chronotope of 'minetime', in his 1946 novel *Mine Boy*, the first fictional narrative of the life of a black miner in South Africa.[26] In the novel, a new recruit is told that if he is to survive in the mines, he must learn to stop fearing that his work may lead to nothing. In time, tomorrow, he is told, he will begin to stop thinking of the future. In time, he is told, he will stop worrying about the fact that the pile of rock at which he labours never seems to grow

26 Peter Abrahams, *Mine Boy* (Oxford: Heinemann, 1989 [1946]).

or diminish. Only when that occurs, advises his mentor, will he be liberated from the fear that possesses him in the mines.

Still, learning to overcome the fear of producing nothing cannot, in the end, save the miner from the other horror, the collapse of the mine in which he and his comrades are swallowed by the earth. And only time will release him from the fear of the violence that is labour in the mines. The 'in-a-file-people', who already resemble the 'corpses forming a line', make 'drill boys [defecate] in their mine overalls', sings Ngoana Tooane Motsoafi, a miner whose song is recorded in David B. Coplan's *In the Time of Cannibals*.[27] But it is perhaps the spectre of the catastrophe—the rock burst, the inundation, the underground fire, or the gas explosion—and its status as the most exceptional of events, that finally banishes the everyday fear of the miner, whose labour constantly disappears, ensuring that the miner remains captive in the cave of another's desire for gold.

27 David B. Coplan, *In the Time of Cannibals: The Word Music of South Africa's Basotho Migrants* (Chicago: University of Chicago Press, 1995).

In Vilakazi's poetry, this captivity makes of the miner something like the metal he seeks, in an awful de-animating mimesis: 'We, too, grow old and rusty in the mines,' he writes, 'Our strength soon goes, our lungs soon rot.'[28] It is a macabre invocation and transformation of those famous lines from the scribe whom Agricola so loved, the poet whose blame of gold for the evils of war he had nonetheless repudiated. Virgil, poet laureate of the shepherd and the farmer, but also the warrior, had understood, as had Agricola, that mining and agriculture are each other's twins, inextricably bound up in the double helix of empire. And this long before the creation of that strange figure, the peasant-proletarian, who would shuttle between country and city, or, as they came to be know in South Africa, the location and the mine.

28 Vilakazi, 'Ezinkomponi'.

Virgil had rendered the relation as kind of reversibility between sickle and sword. In the *Georgics*, the awful predicament of war is described as one in which 'respect for the plough is gone; our lands, robbed of the tillers, lie waste, and curved pruning hooks are forged into straight blades' (1:506–8). Peace is for Virgil a futurity without guarantee, and it can be imagined only from the perspective of the *future* farmer, who will till the soil of plains and 'find javelins corroded with rusty mould, or with his heavy hoe will strike empty helmets, and marvel at gigantic bones in the upturned graves' (1:493–7).

Secreted within that much-cited image, with bronze or steel at its centre, is the recognition that empire is always a project of mining, and that mining is as much the condition of possibility of agriculture, at least of the sort that depends on metal implements, as it is the result of a surplus that sustains it. One might say that mining is both ground and underground, the shadow and the substance of the imperial world. This was already true long before the Romans undertook their deep underground operations in places as remote as Anglesey, in Wales, but most notably at Rio Tinto, in the same site that had been mined by the Etruscans and the Tartessians for 3,000 years before them. Rio Tinto: the red river. Its copper-soaked banks are the same colour of dried blood that you find along the canals and pipelines leading away from the ERMP's mines in Boksburg today. Rio Tinto is also the name of the company that took its name from the Spanish site and later merged with a capital-rich lead and zinc corporation founded by Hoover, originally financed (at least in part) with the proceeds from the recruitment of Chinese workers for the mines on the East Rand.

Duplicate Keys

Reading the Cash Book, one is struck by the accumulation of coincidences, simultaneities and the weave of abstruse threads binding distant worlds and times. Virgil wrote his songs of the farmer and the shepherd in the aftermath of imperial warfare. So, too, British imperialism developed a new pastoral aesthetic, in which the world to be colonized was figured as a field to be harvested and a population to be shepherded, during the twinned period of revolutionary republicanism and industrial revolution. That aesthetic grew out of a pastoralism that had once depended on the opposition between the fantasized labourless existence of the shepherd and the painful labour of the farmer, but that gradually replaced it with a simpler division between the country and the city, as the rural became confused with nature and the Native. In the aftermath of that cataclysmic refiguring of the world, which coincided with British expansionism and its assumption of the place formerly occupied by the Dutch and Portuguese in the maritime world, not least in South Africa, the colonies became the scene and the screen for a nostalgic projection, which would nonetheless capture the world as if in the shadow of the magic lantern. Unfurling in that strange half-light was a narrative of the world progressing from agriculture to industry. It had as its allegorical emblem the company garden, like the one planted with select seeds obtained in Johannesburg.

The plough is accounted in the Cash Book at £2.15, the harrow at £1.10, the rake at £5 and the grass-cutting machine at £3.5 (p. 36v). There is no sword, though the *asegais* of the Zulu insurgents bristle in the margins, and between the lines. There is, however, a gun: a Webley Revolver, and 50 cartridges, bought from one C. H. Smith. According to the Cash Book, the money for the purpose was advanced to S. H. Godwin on 11 December (p. 88r), in the month that the European miners were given their pay for having helped to suppress the Bambatha uprising. In answer to Virgil's sickle and sword, then, the Cash Book gives us a harrow and a handgun: these the alternative destinies of the unprecious metals, paid for with the precious ones.

In the end, however, Agricola, the philosopher of stone, was more interested in Virgil's poem of war, the *Aeneid*. Perhaps the most haunting moment of that verse comes in the image of a frieze, which is to say the image of an image, depicting young Troilus' death. Troilus is the boy, unequal to Achilles, who falls beneath his chariot and is dragged by his hair in the madness of battle. A. S. Kline translates the text in a manner that reminds us of that central reversal upon which the entire body of Virgil's poetry works. The boy's spear 'reversed furrowing the dust';[29] John Dryden rendered it in the poetry of his own time as follows: 'The hostile spear, yet sticking in his wound, / With tracks of blood inscrib'd the dusty ground.'[30] Here, the weapon that has not yet been buried for future farmers to unearth, becomes not a plough but a writing instrument; the stylus is wetted with blood.

The problem is that the death written there can't be read, even when it is accounted for. For death is that which can never be read. This is because, as the most fundamentally Real, it never, as Jacques Lacan says, ceases to not be written. And yet, there is a politics of not writing the deaths of others which exceeds the question of the Real and the Symbolic, the Symbolic being, for Lacan, that which never ceases being written. Is that not what confronts the reader of the Cash Book where

29 Virgil, *The Aeneid* (A. S. Kline trans.), 1.47. Available at: http://www.poetryintranslation.com/PITBR/Latin/VirgilAeneidVI.htm (last accessed on 20 August 2014)

30 Virgil, *The Works of Virgil Translated into English Verse by John Dryden* (London: J. Walker, 1818), p. 138.

she is confronted by a writing at the limit of writing, where story evaporates in number? Death is counted there but never fully accounted for. And where it is only counted, where no name marks the spot of a vanishing, the task of reading becomes a pursuit of what remains, not a forensics of the remains, but an orphic search for some echo wherein a voice might have sounded. Sometimes, though, one is caught in the war of words.

Cain, Interpreter

It is well known that among the many injuries suffered by miners, hearing loss is an especially common one. Explosions, the constant hammering of pick on stone, and the descent and emergence from great depths all take their toll on the ears. The problem is that the prevention of industrial hearing loss often entails covering one's ears, but doing so means also limiting the capacity to hear those other sounds that might foretell a falling rock or an imminent collapse of the tunnel: cracks and rumblings, the wordless language of the earth, the oracles of accident. Unless and until machinery itself is made quiet, miners are always going deaf.

But, and at the same time, mining and especially gold mining has also always been the scene of overhearing, of eavesdropping and of espionage. This is partly because miners have the most immediate access to gold (or diamonds or other precious metals) in the darkest recesses of the earth. It is thus possible for them to steal and to hide small fragments of valuable earth on or in their bodies and clothes. For this reason, the mines have developed elaborate security measures, ranging from the use of body-cavity searches and public showering to metal detection devices and electronic surveillance systems. Even before 1906, they had developed strategies for spying on miners in the mines, forms of eavesdropping. But, of course, to overhear is not yet to understand what is being communicated. The message always requires interpretation.

On 21 August 1906, advances in the amount of £74.10 were paid to one E. Cain, an interpreter, who was sent to China from the East Rand. If there is any name in the Cash Book that commands the philologist's attention it is this one. In the book of Genesis, Cain is the farmer, the first to be born and the first to kill. He is also brother to Abel, the herdsman. It has been argued that the name Cain is related to the Arabic word *qyn*, meaning blacksmith, a worker of metals.[31] If so, it provides a linguistic ground for Agricola's insistence that agriculture and mining are primordially linked, but also for Virgil's insistence that they are tied to violence. Not only because of the love of gold, but because war rests on the industrialization of violence and on the surplus that is at once its product and its sustenance.

31 Richard S. Hess, *Studies in the Personal Names of Genesis 1–11* (Ann Arbor, MI: American Oriental Society, 1993), pp. 24–5.

For centuries there has been an intimacy and even a mutual dependency between mining and military activity. The one is the forge of the other. Imperial warfare has always depended not only on the mining of metallic minerals (as Virgil well knew) and on tempering metal with heat but also on the use of explosives developed in the awful doublet that binds mines and might. In English, the word mine can equally refer to the explosive device or the cavity in which it is placed, and mining, as a verb, may refer either to the excavation of the earth or the planting of those devices that will displace it. In French, as Peter Szendy tells us, the word *écoute* discloses an uncanny history; the verbal form, *éscouter*, which in the late seventeenth century meant 'to attend with one's ears' became, in its substantive form, *l'écoute*, which referred not to listening but rather to the place from which one listened without being seen. This was the French that was brought by the Huguenots to the Cape of Good Hope when they fled persecution following the Edict of Fontainebleu (1685), which had ended the religious toleration so precariously secured by the Edict of Nantes (1598).

By the nineteenth century, the word '*l'écoute*' had come to refer to a place from which one could hear 'without seeing or being seen', and which, more importantly, was defined as a small gallery in a mine where one could hear 'if the enemy miner works or moves'.[32] The enemy miner was sometimes a soldier planting mines beneath the surface of the ground, in trenches or beneath roadways, but he was also, perhaps, the resistant miner, the one whom mining capital always feared would rise up in open warfare against the shift bosses and the mine managers, and the capitalists themselves. The dictionary entry of the *Grand Larousse* makes it impossible to decide. But it was against such enemy miners that the companies of the Rand would form their coherent policies—not only against the white strikers of 1914 but also against Native opposition to

32 Peter Szendy, *Sur Écoute: Esthétique de l'Espionnage* (Paris: Les Éditions Minuit, 2007), pp. 24–5.

colonial rule, as when they willingly paid the salaries of the men who left the mines to fight Bambatha's insurrection.

The redoubled relationship between mining and militarism, which in English resides in the word mine itself, resonates in the language of espionage, where overhearing implies a kind of clandestine capture or interception of messages. But another doubling occurs in Afrikaans, where the word *luister* means both to listen, in its verbal form, and lustre or radiance, in one of its noun forms. In Afrikaans, there is an affinity between that which can be heard and that which shines forth in visibility. Indeed, it is difficult not to think that the word *luister* belongs to gold and is its proper adjective. The radiance of the kings, to invoke the title of Camara Laye's novel, which opens with the image of a young monarch weighted with gold, has always depended on this precious metal.[33] And a false etymology, one that arises from mere sonic proximity (a near rhyme), implies that the secret of gold's value is as much a function of overhearing, as being seen. After all, a mere rumour of gold can start a rush. One is tempted to think that this is why, when describing white colonialists, Sotho speakers (who would form a large part of the mining labour force throughout the twentieth century) remarked the transparency of their skin by saying that the sun could be seen shining even through their ears.

33 Camara Laye, *The Radiance of the King* (James Kirkup trans.) (New York: MacMillan, 1971[1954]).

Whatever the origins of that image in Sotho, Afrikaans, like Latin, allows the two words, one for gold, the other for listening, to confuse themselves in the ear. Thus, on the pages of the Latin dictionary, *aureus*, meaning golden, which implies not only gold but the radiance of the sun, sits next to *auris*, meaning listening. And it should be noted that the guarding of secrets in gold-mining circles, from the results of assay testing to the location of claims, is undertaken with as much rigor, and sometimes as much force, as those of states and other forms of power whose security depends on secrecy.

Telephone Communication

When miners speak, as they often do, of their 'sixth sense', meaning their capacity to hear in advance the signs of an impending event in the earth, they are referring to this kind of overhearing. Rather, they lay claim to discern something which, today, is said to be available only to the most advanced technology. One of those technologies is the fantastically named geophone, a device that listens to the earth, converting sound waves into electrical signals and allowing for the measurement of movement within the earth, namely the sending and receiving of extremely low frequencies. Another, derived from it, is the tactical ground sensor, which, like the nineteenth-century *écoute*, is also an apparatus used for military purposes: it uses acoustic and other information to sense the activities of the enemy (exactly, as the *Grand Larousse* says, his 'work and movements'). Extremely low frequency technologies are also currently being developed as part of an emergency communications apparatus that would be capable of detecting the voices or bodily signs of miners in the aftermath of an accident. It is an ameliorative or supplementary apparatus, a technology addressed to the inevitable failures of other technologies.

High frequencies cannot traverse rock. Like light, which can be easily interrupted, they cannot be heard if the rock obstructs their paths. Still a dream of the future, the technologies that simulate the miners' sixth sense are always, nonetheless, belated. As are all rescue operations, even when they seek to arrive 'just in time', on the thin line where accident can divide the living from the dead. For this reason, one cannot help but think that the geophone is the true heir to Edison's first listening machines, which he (and many others, in many parts of the world) thought would allow him to communicate with the dead. Perhaps, too, the telephone is the true heir to the miners' sixth sense. After all, to speak and hear those that one cannot see, especially when one is underground, can seem like a miracle.

Balance Forward

T. J. Haarhoff (1892–1971), the great South African classicist, who was a devoted reader and interpreter of Virgil, and the ideologue for a humanist Afrikaner nationalism that fancied itself the distant heir to Roman imperial authority via Dutch Roman law, was also a spiritualist. In seances that he later conducted in his house on Rhodes Avenue, in Johannesburg (named for the Kimberley mining magnate), he sought direct inspiration from Virgil on translational matters and found in the ancient poet the author of an original universalism. Like so many other universalisms, however, it was premised on a constitutive exclusion. Thus, Haarhoff, who had travelled to England to study Greek and Latin—on a Rhodes scholarship, despite being of

Afrikaner descent—advocated a rapprochement between Afrikaners and the English, whom he imagined in the image of the farmer and the colonial fighter, respectively. But this confraternity was not to admit Africans, whom he believed required 'strict supervision'.

Thus does the word return. Supervision. Oversight. The view from above. Haarhoff's house overlooked the city from a vantage point that would later enable South African police forces, under the leadership of Jan Smuts, to fire upon striking miners. Haarhoff was an ardent advocate for Smuts, who held three portfolios—as minister of interior, defence and mines — in the first government of the Union of South Africa. Even though he had been ousted from those posts by a restive faction of separatist Afrikaners called the 'Old Boers', Smuts (and Botha) would negotiate the ceasefire that ended the (white) miner's strike in 1914. Smuts would cut his teeth in the war on striking miners in 1907, just a few months after the 1906 Cash Book concluded its register of credits and debts, profit and discontent. At that time, he was not yet minister of mines but a general and colonial secretary of the Transvaal. When white miners rejected the proposal to reduce their numbers vis-à-vis black labourers (or, as they saw it, to increase their supervisory responsibilities) and went on strike, Smuts deployed 1,400 imperial troops on behalf of the mines and cemented an already well-developed tradition that would be renewed repeatedly until August 2012, when police fired on striking miners at the Marikana platinum mine owned by Lonmin, killing 38.

It is in the strange form of a prophecy recovered in retrospect that Marikana makes its appearance on the grieving pages of the Cash Book. At Marikana, it has been said, the South African democratic experiment threatened to run aground. Once again, the state and capital allied themselves against the workers, aided by the trade union's security forces. The conjuring trick demanded in the aftermath of that bloodbath is to rekindle a belief in the possibility of economic justice, there where the workers of the earth first created the scaffolding of the political underground—which, in the apartheid years, often took the form of the National Union of Mineworkers. Today, conditions of life in the informal settlements that crust the compounds and the fenced territories of operations, many of which resemble the technological nightmares of a Fritz Lang film, are not much different than they were in 1906. For many, there is no electricity. No running water. No sanitation. Wages are low and lowest for those whose lives are most endangered by the work and the bonus systems still in place.

There is good reason to believe that it is in the mines of South Africa that the lunatic apotheosis of biopolitics in necropolitics was first dreamt. From the rationalist nightmare in which life was cultivated, nourished and enhanced merely for its instrumental function, there has emerged a new economy liberated even from that bad dream of utility,[34] one based on the glut and disposability of bodies. Marx long ago described the operations of capitalism as necrophiliac, not merely because it destroyed lives and fed on death but because it demanded that labourers become the mere helpmeets of machinery in which was congealed the residue of previous workers' labour as well as the intellectual labour invested in technology. But the necropolitics that has arisen at the scene of mining in South Africa is something else, a kind of abandonment in which increasingly rationalized technologies have produced a superfluity of bodies. These working bodies are no longer prisons of the soul addressed by an educational system and other ideological institutions that imbue them with a love of productivity. In the rural peripheries which send most of the labour for the mines today, literacy is nearly non-existent, and education is minimal.

Foucault once argued that the dream, as a place of doubt, is the other of madness. But you can't think about the gold mines without asking: What terrifyingly reasonable madness dreamt this world?

The commission of inquiry into the events at Marikana was established in August 2012 and has, as of mid-2014, failed to produce a judgement about what happened, not only at the Lonmin platinum mine but also in the days preceding the confrontation, days in which a war between representatives of the National Union of Mineworkers and an emergent organization called the Association of Mineworkers and Construction Union led to the deaths of several miners. The commission is a kind of collective audit of the mining sector more than a century after the EPRM Cash Book was written. Auditing is, etymologically speaking, a hearing, though not necessarily an act of

34 Achille Mbembe, 'Necopolitics' (Libby Meintjies trans.), *Public Culture* 15(1) (2003): 11–40. On biopolitics, see Michel Foucault, *The Birth of Biopolitics: Lectures from the Collége de France, 1978–79* (Michel Senellart ed., Graham Burchell trans.) (New York: Palgrave MacMillan, 2008[2004]).

listening. It is better grasped as listening in the mode of judging. It requires a balancing of the double column, of costs versus revenues. What is left over is ostensibly profit. But the expenditures must be real, and the remainder must be equal to the starting balance less expenditures and amortizations. Otherwise, one suspects malfeasance.

Once again, language is a crypt whence a social logic can be excavated. Amortization: from the Middle English 'amortisen', and derived from the Latin *admortire*, meaning to kill. To kill value, or depreciate it, is what time does. But only because it has been regulated by law, the law that calculates the cost of time for labour and things, and especially capital-intensive technology, such as is used in mining. Law also makes time the source of value for money, in the form of interest. But if, as anthropologists say, this principle of interest is inherent in every society and every history of exchange, if it is what compels reciprocity itself, the rate is not universal. Nor is the sanction that befalls the one who cannot make payments in time. In time, interest grows—how fast or how slow is a matter of fiscal policy and lending practice. But, as interest grows, so does debt, for it is only the promise of debt repaid that makes interest real.

Marx referred to interest-bearing capital as the most fetishistic form of money, the point at which the magic of transubstantiation performed by the commodity form reaches its apogee. Interest, one could say, is the radiance of gold become abstract, but no less powerful for that. And it is for this reason, because of the demonic powers of interest, that the balance forwarded from 1906 may well not be only profit but the events of Marikana and the political crisis that is their aftermath.

Following the strikes and the massacre, it was revealed that at least part of the urgency that afflicted the demands for higher wages, as well as the need behind those demands, was a function of predatory lending. Miners, and especially rock-face drillers, who had taken loans out at exorbitant interest rates, both from the dominant union, the NUM, and microlenders functioning as fronts for larger financial bodies, found themselves increasingly unable to repay these debts. Faced with the demand for money that they did not earn, they struck. But because they undertook their strike off schedule, at a time not permitted according to the calendar of contract negotiations, their strike was simply called a wildcat action, and they became vulnerable to firings, criminal prosecution and even imprisonment.

In the brief two years since the strikes, the balance of power between unions has changed, a new political party has emerged, some mines have closed, others reopened, strikers have been fired and re-hired, men and women have made their way back to the rural peripheries to bury their dead and sing of their losses, as others have returned to the mines and the underground. More strikes have occurred, some stretching across months. Having accused their union leaders of failing to listen, some now accuse the commission of inquiry of being 'both blind and deaf'. The desire to be heard remains the siren song of the disenfranchised.

The East Rand Proprietary Mines Cash Book records no voices. Nor can one speak for its absent subjects, named or unnamed, in the lines and columns that link and separate them. To write them, as in a letter, would be impossible. Nor can any telephonic communications reach across the century. Is my own text, then, a rite, a conjuration of the spirits that inhabit the myths and dreams that made gold mining what it is in South Africa? Perhaps. But if so, it is because that world was itself born of a myth and a dream. What appear to be foreign correspondences turn out to be at the heart of gold mining, the origin of a globality whose involution now produces value at the limit of what can be counted. So: Farrar and Rhodes, Cinderella and Hercules, Hersch and Hoover, Virgil and Agricola, two empires (Roman and British, to be displaced by two more, American and Chinese), a boy with a number and a man named Fifteen, a woman named Bella, migrants Wang Hai and Jong te Jim, Bambatha and the wandering Chinese, Mungul Singh and Mohandas Gandhi, Rio Tinto and the Hoover Dam, a harrow and a handgun, a mine which might have been a war machine, a gold bar and a colour bar, an accident and a crime, a speech overheard, a worker overseen and a catastrophe overlooked, profit and prophecy. In the Cash Book, these correspondences are the invisible presences and contain the truths to be told about the empire of gold in the echo chamber of South Africa's mines.

* A gold mine in South Africa, one of the deepest mines in the world.

JONAH AND THE WHALE AT BLYVOOREIZICHT*

Word
a mouth.
Break open the sky,
make it blue and bluer still:

Methane. Down here the world is
dark as the first day;
we look for light in the canary's wings,
sound the dark with sirens.
We are quiet, for we are
exceeding afraid.

Word
a mouth.
and water like language
rushes in to drown us.

Rock burns, yet oceans beneath the land
are home to tempests. We go down
to possess them, to rage,
vehement as the east wind
that does not blow there.
And the depth closes round about us.

Word
a mouth.
We are as fish, now
breathing without our lungs.

Our lungs are black. For we went down
to the bottoms of the mountains.
The earth with her bars
was about us forever.
We held our breath as one lover
holds to the other on the last day.
We pursed our lips, and
closed our eyes. We went down
and our souls fainted. For we breathed
without our lungs.

Word
formless.
The mouth opens
as does the earth.
And as water rages below,
soil rises, red like a whisper.

We cried by reason of our affliction:
our fear, our black lungs, the violence
which was in our hands. We cried.
Our affliction was our reason.
We cried that innocent blood
not be laid upon us. Our fear, the violence,
that it might be a shadow above our heads.

Mouth
                    a mouth.
(just one?)
Let it be   rib.

Broken. This is all the eye beholds:
brace of steel or arc of bone,
unbleached, holding the earth up,
the heart in place.

Rib
                    sceptre
tooth and bone as relic
if the body is not yet burnt.

We hold our breath, as
one lover holds to the other
on the last day. Now,
save us from temptation,
hold our hearts in place.

Jonah,
                    faithless
with limbs aching, swims
against the current.

We try to recall a moment
before this, before we cast our lot
and the lot fell upon us. We
who have doubted, who have
only doubt, living our breathless
lives in amber oblivion, we
try to recall
                    but are dumb.

Ear:
  a tin drum.
Strike and strike again,
teach us the sound of wailing.

Teach us the sound of wailing,
and lay not upon us
innocent blood, but fill our mouths
with hellfire and righteousness
that we may know,
before the plague comes,
how to save ourselves.

Belief:
    whale
in a land-locked land.

The earth with her bars
is about us. Down, down
beneath the bottoms of the mountains
we are afraid, our souls faint,
and a shadow above us falls.

Jonah,
    faithless
with limbs aching, swims
against the current.

Because we were impatient,
because the shade was not enough
and the heat was far too great,
because the desert at least
lies beneath a blue sky,
we bloodied ourselves, and threw
flesh upon the pyre, and prayed
god receive our burnt black offering.

Go down again,
    Jonah,
go down
to that place

where the rock burns
and the air is wet enough
    for fish.

Jonah,
from his ashes,
spake not a word. Yet, we bid him
descend once more.

Go down, Jonah, go down
into the deep, go down
to the mountains' end,
let the earth close round about you
for there is gold
in the tooth of a whale,
Jonah, there is gold
in the tooth of a whale.

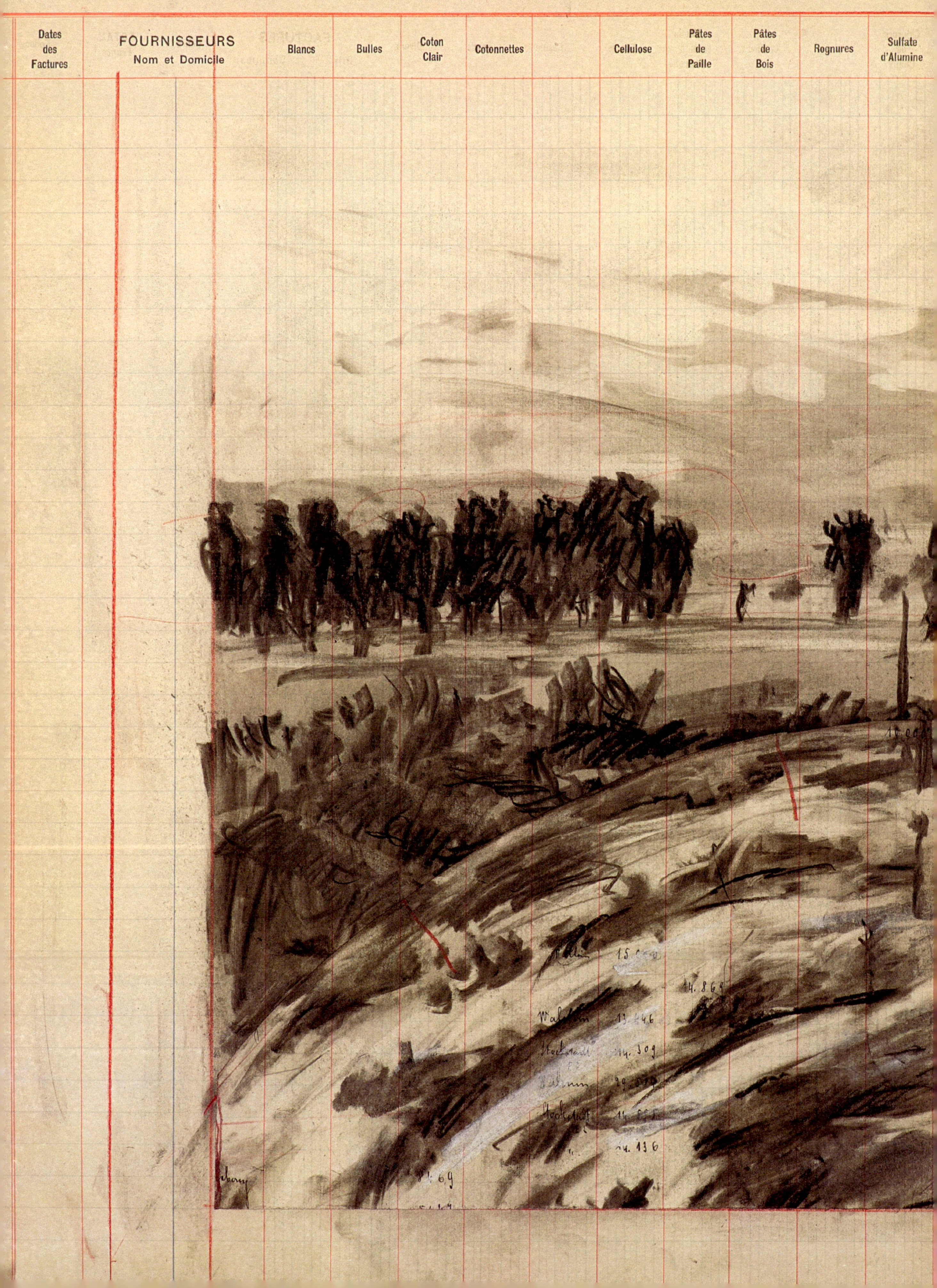
Dates des Factures
FOURNISSEURS
Nom et Domicile
Blancs
Bulles
Coton Clair
Cotonnettes
Cellulose
Pâtes de Paille
Pâtes de Bois
Rognures
Sulfate d'Alumine

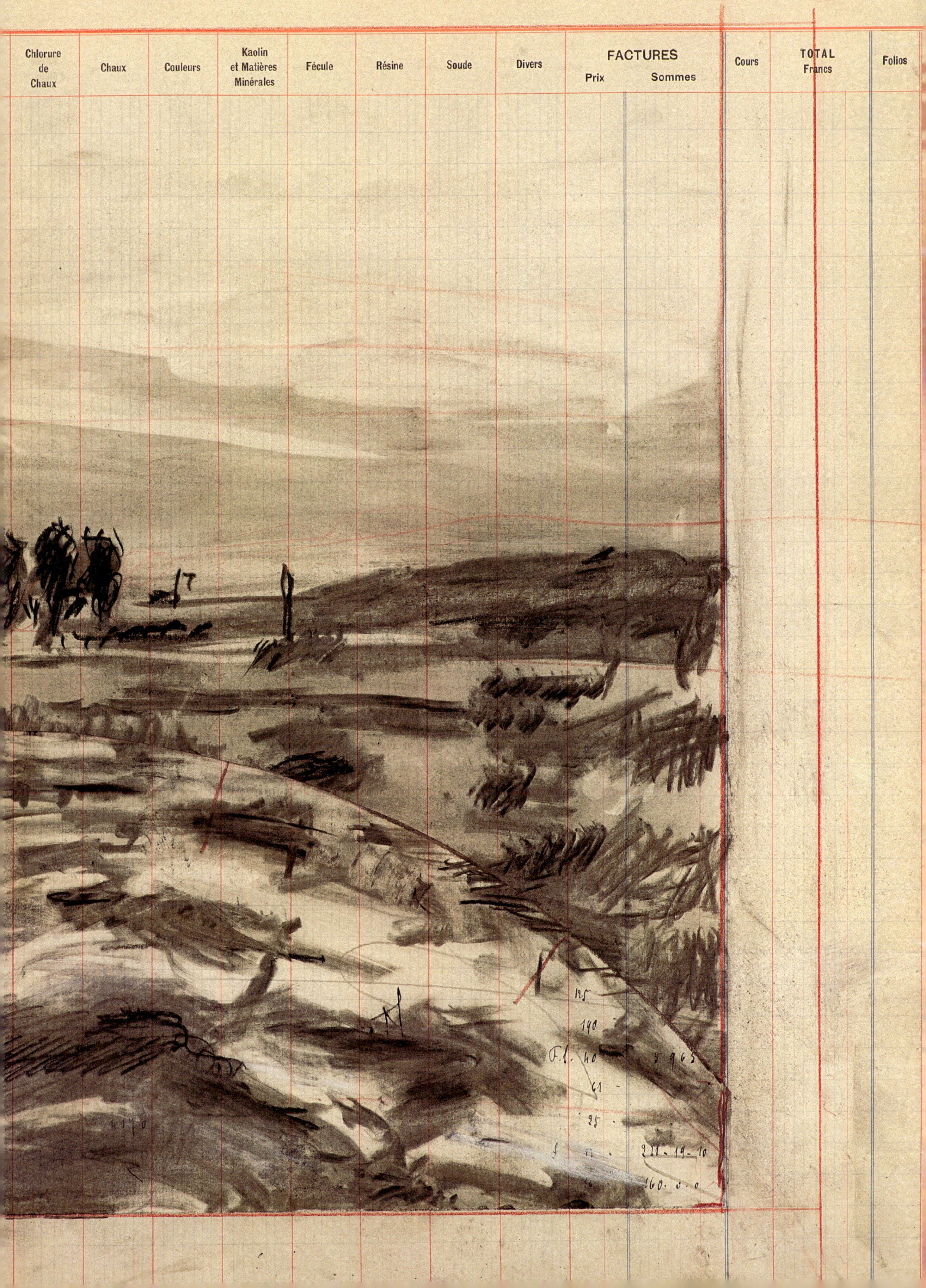
Chlorure de Chaux
Chaux
Couleurs
Kaolin et Matières Minérales
Fécule
Résine
Soude
Divers
FACTURES
Prix
Sommes
Cours
TOTAL Francs
Folios

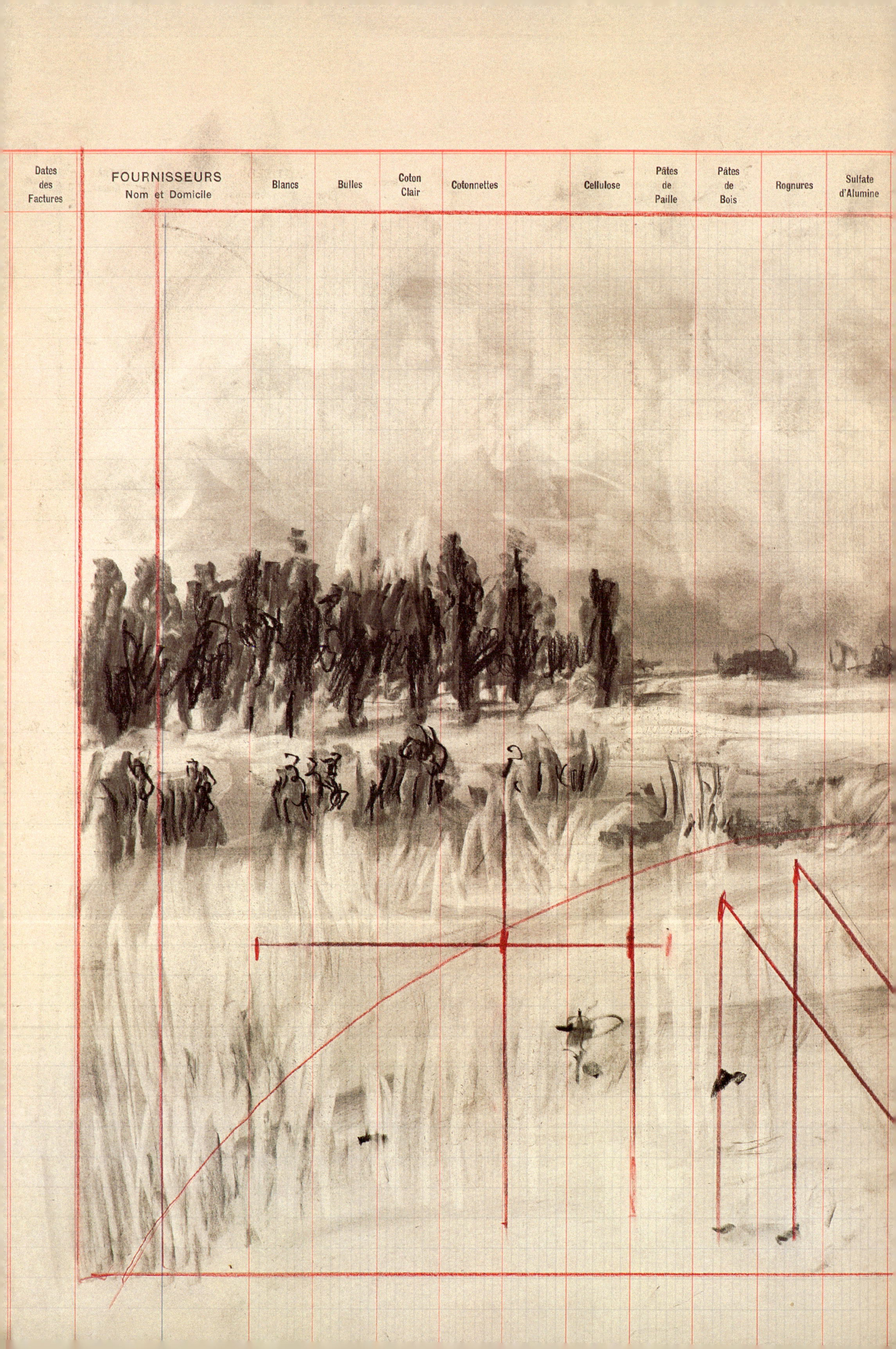
Dates des Factures
FOURNISSEURS
Nom et Domicile
Blancs
Bulles
Coton Clair
Cotonnettes
Cellulose
Pâtes de Paille
Pâtes de Bois
Rognures
Sulfate d'Alumine

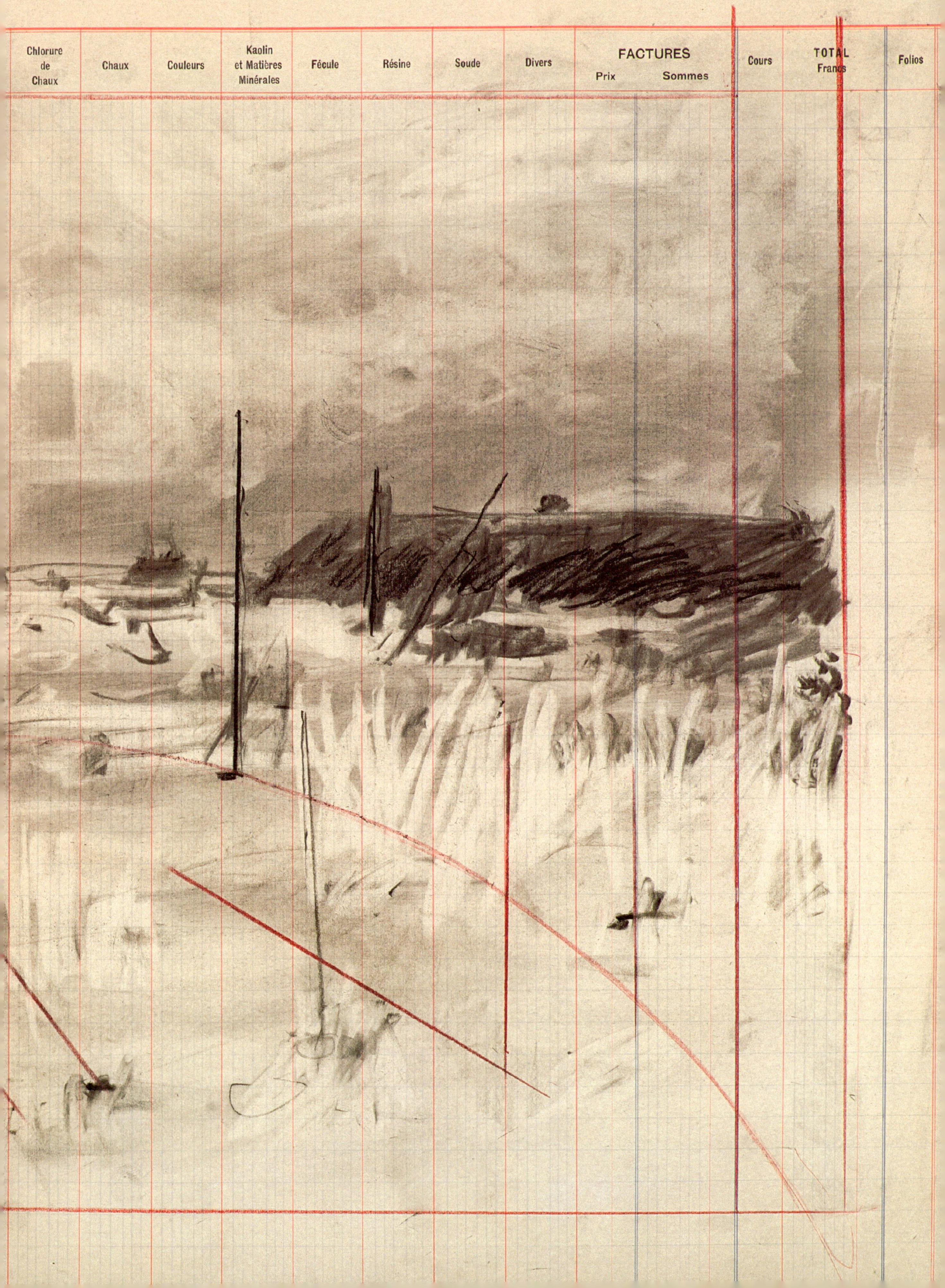
Chlorure de Chaux
Chaux
Couleurs
Kaolin et Matières Minérales
Fécule
Résine
Soude
Divers
FACTURES
Prix
Sommes
Cours
TOTAL
Folios

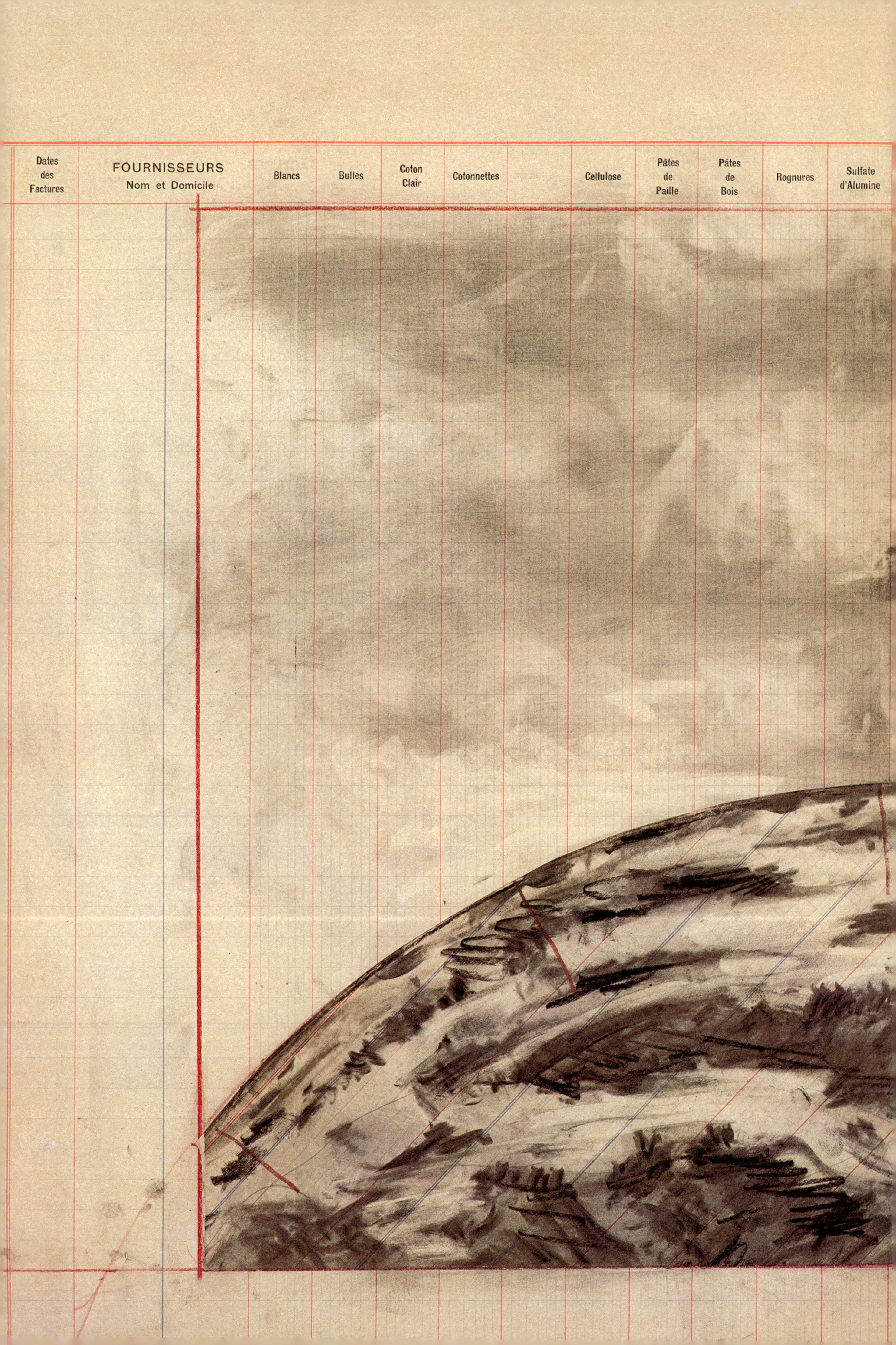
Dates des Factures
FOURNISSEURS
Nom et Domicile
Blancs
Bulles
Coton Clair
Cotonnettes
Cellulose
Pâtes de Paille
Pâtes de Bois
Rognures
Sulfate d'Alumine

Chlorure de Chaux
Chaux
Couleurs
Kaolin et Matières Minérales
Fécule
Résine
Soude
Divers
FACTURES
Prix
Sommes
Cours
TOTAL Francs
Folios

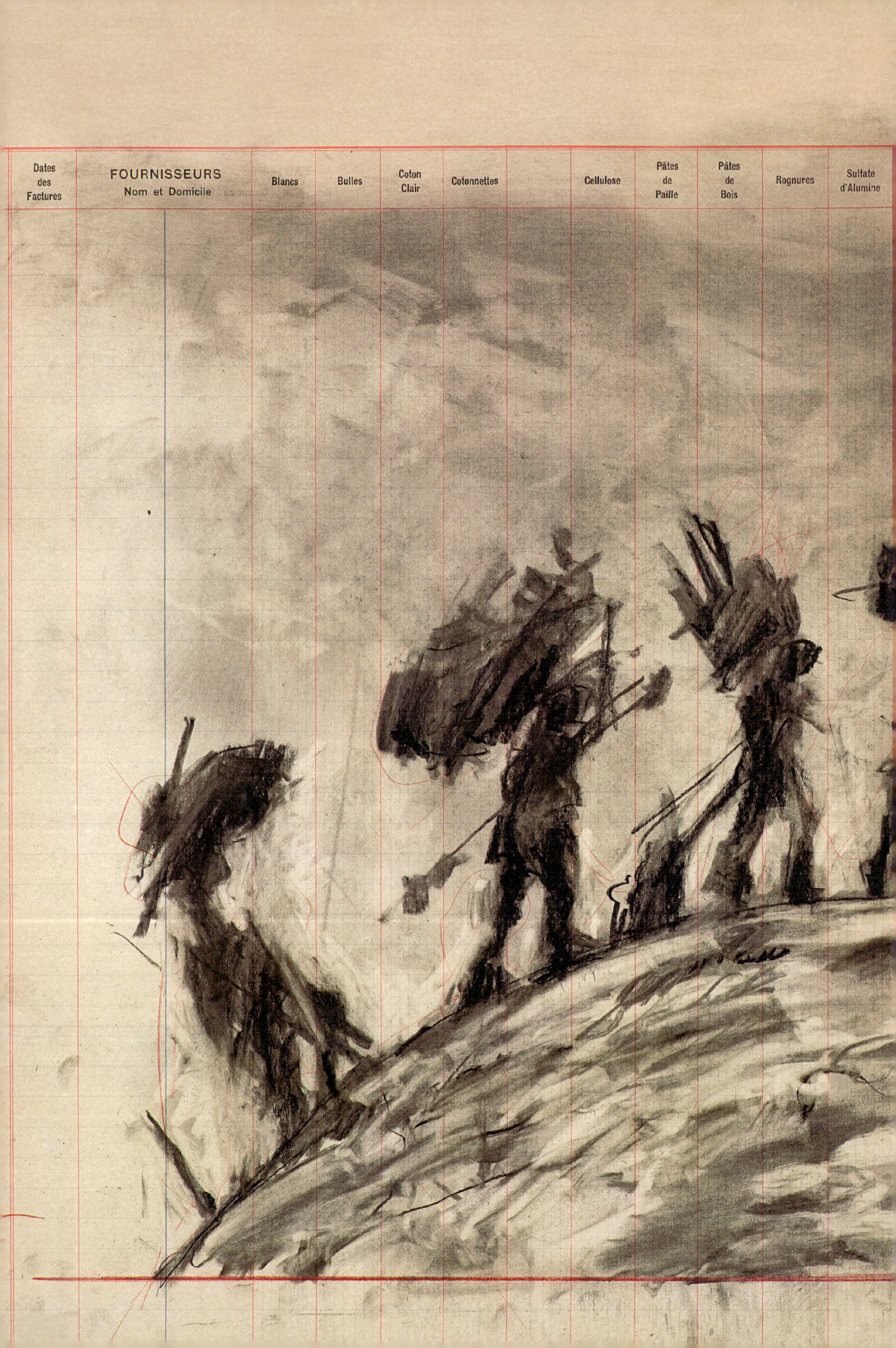

Dates des Factures
FOURNISSEURS
Nom et Domicile
Blancs
Bulles
Coton Clair
Cotonnettes
Cellulose
Pâtes de Paille
Pâtes de Bois
Rognures
Sulfate d'Alumine

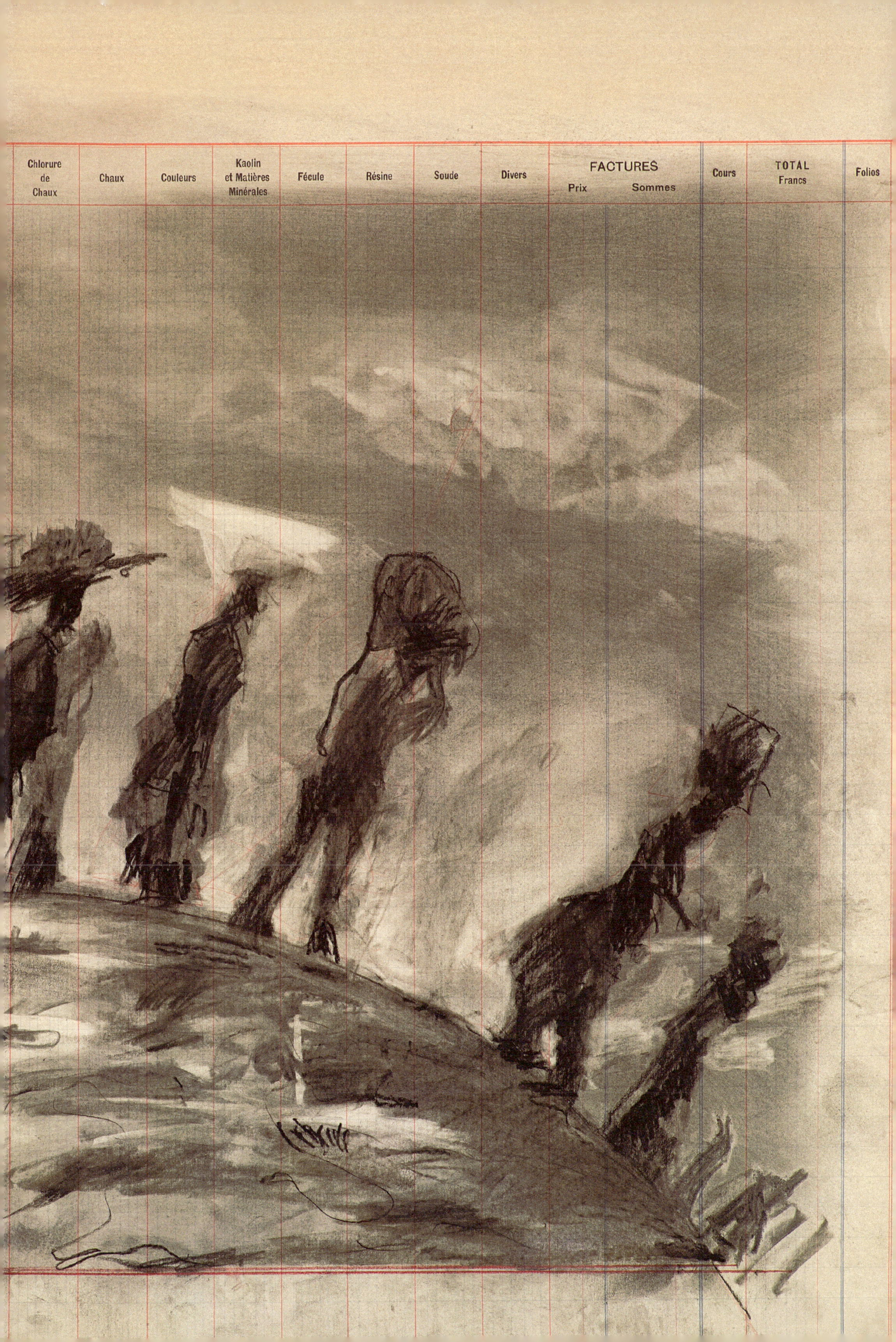
Chlorure de Chaux
Chaux
Couleurs
Kaolin et Matières Minérales
Fécule
Résine
Soude
Divers
FACTURES
Prix
Sommes
Cours
TOTAL Francs
Folios

Dates des Factures	FOURNISSEURS Nom et Domicile		Blancs	Bulles	Coton Clair	Cotonnettes		Cellulose	Pâtes de Paille	Pâtes de Bois	Rognures	Sulfate d'Alumine
1933												
6 Nov.	Falchenberg. Musin	Paris					Hallein	15.041				
3 Oct.	Stichstoff Fabrik	Veendam							15.988			
16 "	Phoenix H. Mahler	Karlsruhe		10.475								
7 "	Cie Prod. chim.	Paris										15.000
30 "	F. Schacht	Braunschweig										
6 "	E. Muvalle, Degroote	Fives-Lille										
"	"	"										
4 "	Amid. & Rizerie de France	Marquette-lez-Lille										
11 "	Jonas Fritz	Benfeld		392	386							
25 "	"			Emballage 502								
26 "	"			4370								
18 "	Th. Welter	Gimmeldingen										
24 "	Ad. mines de	Bauwiller										
26	Prod. chim. Morosage	Prod. chim. Morosage Mulhouse										
18 "	Talcs	Luzenac										
25 "	Edgard Maillet	Paris								1070	
5 "	A. Sorg & Cie	Martigny									10.415	
20 "	Etabl. Kuhlmann	Paris										
21 "	"	"										
"	"	"										
9 Nov.	"	"										
21 Oct.	Socan. de mat. color. & Prod. chim.	St. Denis										
"	"	"										
7 Nov.	Prague Neuviedl	Prague					Rattimau II	14.777				
18 "	Socan. p. la fab. de la Cell.	Bratislava					Zilina Zvenbl.	15.060				
14 "	Zellstoff Syndikat	Berlin					Mitischalich	15.562				
24 "	"	"					E.Z.E Riedenfelden	15.055				
28 "	"	"					Redenfelden	15.119				
13 "	Falchenberg. Musin	Paris					Hallein	10.049				
20 "	"	"					"	14.887				
25 "	"	"					"	10.047				
10 "	Hennebault	Lomme-lez-Lille				Vichase 11.311						
4 "	"	"				10.540						
4 "	Amidonnerie & Rizerie de France											
10 "	G. Dennis & Cie	Bordeaux										
15	Brinder & Cie	Arches										

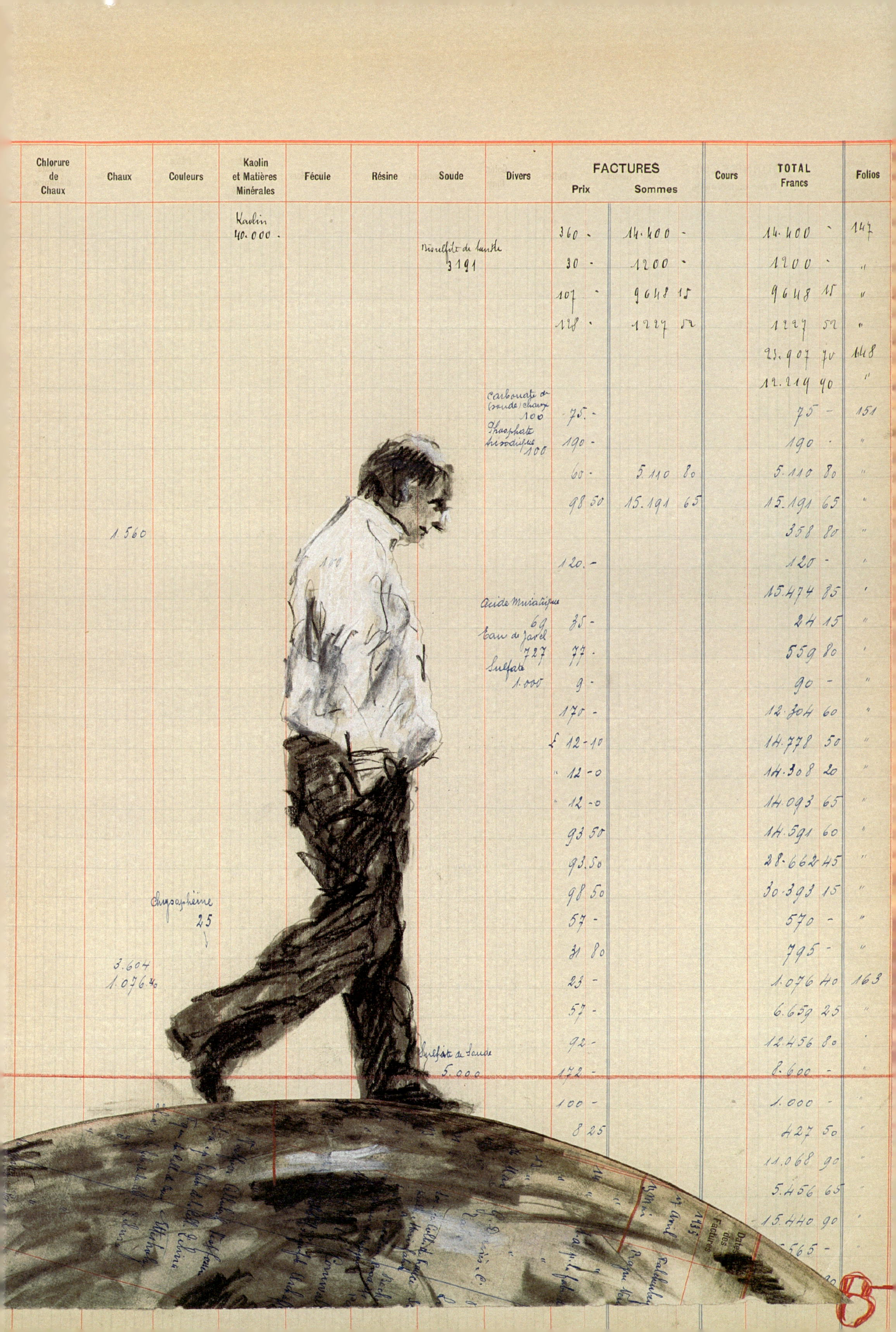
Chlorure de Chaux
Chaux
Couleurs
Kaolin et Matières Minérales
Fécule
Résine
Soude
Divers
FACTURES
Prix
Sommes
Cours
TOTAL Francs
Folios

FOR THE MONTH OF ____________ 190

DATE.		Ledger Folio.	General Charges.	Native Labour Commission.	Native Wages.	Chinese Wages.	Sundry Persons.	General.
	Forward							

CENTRAL ADMINISTRATION MINE CASH

DATE.		Ledger Folio.	Sundry Persons.	East Rand Proprietary Mines.	Native Advances.	Chinese Advances.	Employees' Accident Assurance Claims Account.	General
	Forward							